CAMBRIDGE LIBRARY COLLECTION

Books of enduring scholarly value

Earth Sciences

In the nineteenth century, geology emerged as a distinct academic discipline. It pointed the way towards the theory of evolution, as scientists including Gideon Mantell, Adam Sedgwick, Charles Lyell and Roderick Murchison began to use the evidence of minerals, rock formations and fossils to demonstrate that the earth was older by millions of years than the conventional, Bible-based wisdom had supposed. They argued convincingly that the climate, flora and fauna of the distant past could be deduced from geological evidence. Volcanic activity, the formation of mountains, and the action of glaciers and rivers, tides and ocean currents also became better understood. This series includes landmark publications by pioneers of the modern earth sciences, who advanced the scientific understanding of our planet and the processes by which it is constantly re-shaped.

The History and Philosophy of Earthquakes

The aftershocks of the devastating Lisbon earthquake of 1755 were not only physical: the scientific investigations undertaken in its wake formed the basis of the science of seismology. Published in 1757, the present work is, in the words of its presumed editor, John Bevis (1695–1771), 'a repertory of all that has been written of earthquakes and their causes', and includes several recent papers published by the Royal Society. At the time, scientists suggested subterranean fires or electrical shocks in the atmosphere as possible causes of earthquakes. This reissue also incorporates a brief 1760 work by John Michell (1724/5–93), which uses Bevis' collection as a source and suggests that earthquakes were caused by seismic waves through the earth: it was one of the first to propose that tsunamis were the result of undersea earthquakes. Both these works rank as important steps in the developing understanding of one of nature's most destructive phenomena.

Cambridge University Press has long been a pioneer in the reissuing of out-of-print titles from its own backlist, producing digital reprints of books that are still sought after by scholars and students but could not be reprinted economically using traditional technology. The Cambridge Library Collection extends this activity to a wider range of books which are still of importance to researchers and professionals, either for the source material they contain, or as landmarks in the history of their academic discipline.

Drawing from the world-renowned collections in the Cambridge University Library and other partner libraries, and guided by the advice of experts in each subject area, Cambridge University Press is using state-of-the-art scanning machines in its own Printing House to capture the content of each book selected for inclusion. The files are processed to give a consistently clear, crisp image, and the books finished to the high quality standard for which the Press is recognised around the world. The latest print-on-demand technology ensures that the books will remain available indefinitely, and that orders for single or multiple copies can quickly be supplied.

The Cambridge Library Collection brings back to life books of enduring scholarly value (including out-of-copyright works originally issued by other publishers) across a wide range of disciplines in the humanities and social sciences and in science and technology.

The History and Philosophy of Earthquakes

Edited by a Member
of the Royal Academy of Berlin

CAMBRIDGE UNIVERSITY PRESS

Cambridge, New York, Melbourne, Madrid, Cape Town,
Singapore, São Paolo, Delhi, Mexico City

Published in the United States of America by Cambridge University Press, New York

www.cambridge.org
Information on this title: www.cambridge.org/9781108059909

This edition first published 1757
This digitally printed version 2013

ISBN 978-1-108-05990-9 Paperback

THE
HISTORY *and* PHILOSOPHY
OF
EARTHQUAKES,

FROM THE

Remoteſt to the preſent Times:

Collected

From the beſt Writers on the Subject.

With a particular account of

The Phænomena of the great one of NOVEMBER the 1ſt 1755, in various Parts of the Globe.

By a Member of the ROYAL-ACADEMY of *Berlin*.

Philoſophiæ genus empiricum quod in paucorum experimentorum anguſtiis et obſcuritate fundatum eſt.---Tum vero de ſcientiarum progreſſu bene fundabitur, quum in hiſtoriam naturalem recipientur et aggregabuntur complura experimenta, et obſervationes, *quæ in ſe nullius ſunt uſus, ſed ad inventionem cauſarum et axiomatum faciunt.* VERULAM. Nov. Organ.

A moſt general help to diſcovery in all kinds of philoſophical inquiry is, to attempt to compare the working of nature, in that particular which is under examination, to as many various mechanical and intelligible ways of operations, as the mind is furniſhed with.

Dr. HOOKE's *Method of improving Natural Philoſophy.*

LONDON,

Printed for J. NOURSE over-againſt *Katherine-ſtreet* in the *Strand*, MDCCLVII.

PREFACE.

THE memorable earthquake which ſpread deſolation along the *Atlantic* coaſt in 1755, and the late frequency of ſuch commotions, in a leſſer degree, all over *Europe*, put the editor of theſe ſheets upon exhibiting ſuccinct accounts of the like events in paſt times, with the ſentiments of the beſt naturaliſts as to their cauſes: In the courſe whereof he has retained entirely the facts, arguments and concluſions of the authors from whence he has extracted his collections, and that almoſt in their own words; without ever preſuming to criticiſe any hypotheſis, much leſs to obtrude one of his own.

Thus, he hopes, he has furniſhed a repertory of all that has been written of earthquakes and their cauſes, to be read over at leiſure, or readily conſulted, by the help of a very copious index.

In the annexed account of the laſt great earthquake he has choſen a kind of alphabetical arangement, for the eaſier turning to its phænomena in particular places; all which, he has very carefully collected from the Philoſophical Tranſactions of the Royal Society, and other litterary memoirs and authentic vouchers; and which, as our very ſagacious Dr. HOOKE rightly obſerves, *ſhould ever be regiſtred as ſoon as the obſervations occur; becauſe of the frailty of the memory, and the great ſignificancy there may be in ſome of the meaneſt and ſmalleſt circumſtances.*

A LIST

A LIST of the ſeveral PIECES from whence theſe COLLECTIONS are extracted.

Added by the Editor,

A Me-

A

Methodical Account

OF

EARTHQUAKES.

PHÆNOMENA, *or* FACTS.

Phæn. I. ON the 7th of *July* 1686 about day-break, between two and three in the morning, a great part of *Germany* and the neighbouring parts of *Italy* felt a tremulous commotion. At *Altorff* and the nearest towns of *Bavaria* and *Suevia*, *Ratisbon*, *Memmingen*, *Nordlingen*, with many others, the inhabitants were awakened out of their sleep, and grievously terrified by the rocking of their beds and jarring of their windows. In other places, as *Inspruck* and *Venice*, the tottering edifices threatened immediate destruction: And at *Hall* the walls, with many towers and stately buildings were shattered, and several of the inhabitants buried or oppress'd in the ruins; the

con-

conſternation cauſing moſt of the reſt to betake themſelves to the open fields, where they continued wandering about for ſome days, under the moſt terrible apprehenſions.

A diſmal and horrible phænomenon of nature this! though not unfrequent at other times and places; and therefore highly deſerving the conſideration of natural philoſophers, in order to inveſtigate its true cauſes.

May we not juſtly exclaim with the eloquent *Seneca*, [a] "When the world is ſhaken, and the "ſolid parts of it drop aſunder, when the fixed "baſes of the rocks are rooted up, where can "we hide our heads in ſafety? Where fly for "refuge, when the globe is falling to pieces? If "the ſtage which ſupports us, and on which ci- "ties are erected, gives away, what can admi- "niſter help? Or how can comfort be found "where our fears oppoſe our flight? Walls may "repel an enemy, and lofty towers ſtop the pro- "greſs even of armies: Havens may afford ſuc- "cour in a tempeſt, and houſes ſhelter from "ſtorms and wind: Conflagrations overtake not "the haſte of thoſe that fly them: Subterrane- "ous vaults and caverns can ſecure againſt thun- "der and lightning, a ſmall quantity of earth "bring proof againſt this celeſtial fire, and whole "countries were never ruin'd by it: A peſtilence "may deſtroy the citizens, yet leaves the city "ſtanding: But an earthquake is a wide-waſt- "ing, implacable, unavoidable calamity!"

[a] Lib. vi. quæſt. nat. cap. 1.

Phæn. II.

Phæn. II. That a natural earthquake never extended over the whole globe, is according to *Stobæus*[b], an obſervation of *Plato*, which *Ariſtotle* alſo aſſerts in very ſignificant terms[c]. The ſame thing is remarked by *Metrodorus*, and other ancient philoſophers mentioned by *Plutarch*[d], and *Seneca*,[e] who at the ſame time explode the opinion of *Thales*, and with reaſon; that the earth may be liable to fluctuations, becauſe it ſwims in water, and that thoſe are earthquakes.

Seneca's words are, "If the waters ſupported the "earth, it would be liable to univerſal concuſſi-"ons, and it would be a greater wonder that it "ſhould ever be at reſt, than if it were perpe-"tually in motion." Sure enough it muſt be ſhock'd throughout, and not in any part alone; for no ſhip can be toſſed by halves. We conclude then, that there is no ſuch thing as an univerſal earthquake, but that they are all particular or partial.

Phæn. III. As to the difference of earthquakes happening at different times, or of one and the ſame with regard to various places; at ſome times, and in ſome particular places, they occaſion a latitudinal and, in a manner, horizontal trembling

[b] Eclog. phyſ. cap. 1. *Nullo recenſitorum ibi modorum moveri terram* ſtatuit, *ſed* ἐν τῷ πανταχόθεν ἰσοτάτῳ κειμένην μένειν ἀκίνητον. τόπους δὲ αὐτῆς κατ' ἀραιότητα σαλεύεσθαι. h. e. *Eam in æquabiliſſimo undequaque loco poſitam immotam manere, loca autem ejus aliqua rariora concuti.*

[c] Lib. ii. meteor cap. 47. κατὰ μέρος δὲ γίγονται οἱ σεισμοὶ τῆς γῆς, καὶ πολλάκις ἐπὶ μικρὸν τόπον.

[d] Lib. iii. de placit. cap. 15. [e] Nat. quæſt. lib. vi. cap. 6.

in ſome particular part of the earth, and its incumbent cities and buildings, with a certain degree of concuſſion or ſhock, which, by a peculiar name, *Ariſtotle* calls τρόμον, and *Seneca, tremor.* Sometimes and in certain places, the impetus is impreſs'd upwards, rather in a perpendicular direction. *Ariſtotle* calls it σφύγμος, or *Pulſus*, and *Seneca, ſuccuſſion.* This makes the earth to rock, like a ſhip at ſea, which *Seneca* calls *inclinatio*, and *Garcæus*, from *Pliny*, *arietatio*, eſpecially when the inclination is from ſide to ſide; and then it is alſo named ἐπικλίντης, *inclinator.* In all theſe caſes whole buildings, and even cities are frequently ſubverted; and ſometimes, eſpecially in the ſecond caſe, the earth is violently burſt aſunder (ῥήκτης) or projected aloft, (βράστης) and according to *Ammianus Marcellinus*, *Braſmutias*, or collapſes inwards, the χασματίας of *Marcellinus*, and the *labes*, *ruina*, &c. of others.

Phæn. IV. Theſe diſtinctions are to be found in *Seneca*[f], and *Pliny*[g], who likewiſe give their names[h]. As alſo does *Ammianus Marcellinus*[i]. The earthquake we mentioned, *Phæn.* I. affords an example of theſe varieties. Here at *Altorf*, and in the neighbouring parts, we found the *tremor:* At *Venice*, *Inſpruck*, &c. they felt the *pulſe*, or *ſuccuſſion*; at *Hall* the *ſubverſion*. *Gaſſendus* takes notice of one wherein nothing but a *tremor* was ſenſible, on the 13th of *January* 1617. On the

[f] Lib. vi. quæſt. nat. cap. 4.

[g] Lib. ii. hiſt. nat. cap. 80.

[h] *Senec.* cap. 21. *Plin.* cap. 82.

[i] Lib. xvii. cap. 13.

6th

6th of *April* 1580, all the *Low Countries* were ſhaken with a *ſuccuſſion* which was felt as far as *Paris*, and *York* in *England*: And the town of *Artric* was rocked to that degree, that ſtones were forc'd out of the walls of towers and churches [k]. *Gaſpar Schottus* was at *Rome* when another happened there in 1654 [l]. The ſymptoms of the *inclination*, and the *arietation* are deſcribed by *Seneca* [m], and *Pliny* [n], which latter gives in the ſame place an account of the claſhing together of two huge mountains with a moſt horrible noiſe, and of their receding aſunder again: And the former relates a thing very ſtrange, of the parting of the ſquare marble ſtones in the pavement of a bath, through whoſe interſtices quantities of water iſſued and returned, and of their ſettling in cloſe order again. The ſame authors give many inſtances of *ſubverſions* and *ruins*; as at *Nicomedia* in *Bithynia*, where a vaſt number of perſons were buried under fallen edifices [o]. *Garcæus* [p] gives the names of twelve cities of *Aſia*, which *Seneca* [q] and *Pliny* [r] relate to have been ſubverted in one night, in the reign of *Tiberius*: *Tacitus* [ſ] affirms the ſame, with this addition, that thoſe who attempted to eſcape into the fields, the gaping earth ſwallowed up, and that whole mountains quite ſubſided, and new ones aroſe out of the plains: We read in *Seneca* [t] of a *commotion* throughout *Campania*, which

[k] *Meterranus*. lib. x. [l] Mechan. hydr. p. 62. [m] Lib. vi. cap. 31. [n] Lib. ii. cap. 82. [o] *Ammian. Marcellin.* lib. xvii. cap. 13. [p] Meteor. p. 304. [q] Cap. i. [r] Lib. ii. cap. 84. [ſ] Lib. ii. annal. [t] Lib. ſupr. cit.

ſhook down ſeveral towns about *Naples. Johnſton* [u] tranſcribes *Cambden*'s account of a miſerable deſolation which happened in *England* in 1571, on the 21ſt of *March: Gaſſendus* [w] deſcribes, from *Fernerius* [x]; the memorable ſtroke given, in one quarter of an hour, to all the towns, mountains and rivers near *Lima* in *Peru*, on the 25th of *November* 1604: And laſtly, *Athanaſius Kircher* [y] affirms that he was an eye-witneſs, not without great peril to himſelf, of the ſad diſaſter which befell the fine town of *Euphemia* in *Calabria*, being ſunk as it were in the twinkling of an eye, and covered over with a lake of ſtinking water, the latter end of *March* 1638; who adds that earthquakes ravaged up and down for fourteen days together about that time.

Phæn. V. After theſe inſtances of paſt times, it may be proper to give a ſuccinct account of ſome late ones, out of my collections at large, from the moſt approved *Dutch*, *French*, *Italian* and *German* writers. The *Rimini* gazettes related that on the 18th of *April* 1662, during divine ſervice, a terrible earthquake threw down twelve churches, and ſhattered other parts of that city; that it continued 'till the next *Saturday* and *Sunday*, whereby thirty one palaces and publick edifices were demoliſhed, and above 700 perſons killed, beſides many more ſadly maimed; and that the neighbouring cities of *Faro*, *Peſaro*, *Sini-*

[u] Admir. meteor. cap. 7.
[w] Animadverſ. in Diog. Laert. x. p. 1049.
[x] Hydrog. lib. xv. cap. 18.
[y] Mund. ſubterran. lib. ii.

gaglia,

gaglia, &c. were not without a ſhare of the calamity. The *Journal des Sçavans* for the month of *May* 1678, mentions a terrible earthquake which began *February* the 5th 1663, about half an hour after five in the evening, and raged throughout all *Canada* 'till *July* following, tho' but for a quarter or half an hour together, almoſt every day or night. Its effects were horrible, as mountains claſhing together, and tumbling partly into the river St. *Lawrence*, and partly removed to vaſt diſtances with their trees ſtanding upon them. Letters from *Cornelius Frank*, preſident and counſellor at *Ternate*[z], to *William Maetſuyker*, counſellor at *Banda*[a], dated *Auguſt* 22, 1673. make mention of two unheard of miracles; the one of the burſting aſunder and diſperſion of the very high mountain *Gammacnorra*, with a violent earthquake, and ſo prodigious an ejection of aſhes, that on the 21ſt of *May*, being *Whit-Sunday*, the air became thereby ſo darkened, that people could ſcarcely diſcern one another: The other of a ſecond and moſt ſtupendous earthquake which the inhabitants of *Ternate* were ſurprized with in the night of the enſuing *Auguſt*, about a quarter of an hour after eleven: It ſplit the mountain of *Ternate* quite from the bottom to the top on the ſouth ſide, and levelled the ſtrong palace of King *Mandarſabas* with the ground. At the ſame time the ſea raged ſo furiouſly, that all the veſſels in the port were in the utmoſt danger of being loſt, and the ſhocks were ſtill violent

[z] One of the *Molucca* iſlands.

[a] Another iſland in the *Indian* ſea.

on the firſt of *September*, when other letters came away. An *Italian* letter of *Antonio Bulifon*, to the captain general of the kingdom of *Sicily*, contains a narrative of an earthquake at *Naples* on *Whitſun Eve*, *June* the 5th, 1688, ſo powerful that it ſhook even the foundations of that city. The houſes at firſt ſeemed to be lifted up, and then inſtantly were rocked backwards and forwards with inconceivable violence, and to that degree, that in ſome towns the bells rang of themſelves; that particularly belonging to the clock of St. *Angelo*, was thrown a full palm out of its gudgeon. What greatly augmented the conſternation was a horrible rumbling all the while, as if the world were turning upſide down. In the month of *June* 1690 news arrived from the iſland of St. *Chriſtopher* in *America*, and likewiſe from *Charles Town*, of ſeveral ſtone houſes being overſet by an earthquake, and then ſwallow'd up; in ſome places, of the earth riſing up in large hills, and of the ſinking of trees into chaſms 7 or 8 feet wide in others. The *Jeſuits College*, and all other free ſtone buildings in St. *Chriſtophers* were razed to the ground. Letters from *Naples* and *Rome* of the 3d and 7th of *February* 1693, brought advice of the ruin of the cities of *Catanea*, *Agoſta*, and *Syracuſe*, in *Sicily*; alſo of *Reggio*, and ſeveral other places in *Calabria*; and that as to the reſt of *Sicily*, near one half was overturned, above 100000 ſouls being loſt under the ruins of no leſs than 27 great towns. That at *Agoſta*, *Taormino*, *Syracuſe* and *Catanea*, there are ſcarce any marks of the walls and fortifications to be ſeen, in which laſt city alone, at leaſt 18000 perſons periſhed;

and

and that the head of the neighbouring mountain, at leaſt 600 feet high, ſunk within its hollow, and left a gap ſix *Italian* miles broad.

Phæn. VI. Theſe ſhocks and burſtings of the earth are accompanied with moſt hideous craſhes and bellowings, called by the author of the book *de Mundo* μυκητίαι σεισμοὶ, and by *Ammianus Marcellinus*, *Mycematiæ*: The like noiſes alſo frequently precede a ſhock, and have been known to happen even when no ſenſible commotion followed. *Pliny* ſays, "They are preceded or accompanied with a diſmal ſound, which ſometimes reſembles the lowings of cattle, ſometimes the outcries of men, and at others, the din of claſhing arms [b]." And *Ariſtotle* gives the like account, adding, with *Pliny*, ὅτι καὶ ἄνευ σεισμῶν, ἤδη πȣ γεγόνασιν ὑπὸ τὴν γῆν [c]. *Veſuvius*, *Ætna*, and *Hecla* confirm this; the laſt of which is ſaid to utter ſuch a plaintive kind of ſounds, that many of the credulous inhabitants take them for the doleful wailings of wicked ſinners in hell. During the 11 days earthquake in *Sicily* in the year 1537, the whole iſland was perpetually alarmed with horrible bellowings, and claps reſembling the diſcharge of large ordnance [d]; and *Kircher* affirms the like of *Calabria* [e].

Phæn. VII. Through theſe chaſms and rendings of the earth, it is no uncommon thing for flames and ſmoaky exhalations to aſcend, and diſperſe themſelves to conſiderable diſtances; and

[b] Lib. ii. cap. 80. [c] Lib. ii. meteor. t. 46. [d] *Varen.* lib. i. geograph. cap. 10. prop. 5. [e] Loco. ſupr. cit.

with

with them ſtones, and torrents of a kind of melted metal are often ejected. Sometimes theſe are fore-runners of the ſhock, and they frequently continue after it, eſpecially from the mouths of volcano's[f]. *Tacitus* ſpeaking of the great earthquake which happened in the reign of *Tiberius*, remarks *effulſiſſe inter ruinas ignes*[g]. So in the earthquake which we ſaid raged eleven days together in *Sicily*, the earth opened with a mighty chaſm, from whence fire and flames iſſued with ſuch violence, that every thing within the diſtance of five leagues from *Ætna* was totally burnt up in the ſpace of four days: A ſhort time after which the baſin threw out an inconceivable quantity of fire, ſparks and aſhes[h]. *Ariſtotle* produces ſome examples of ancient times[i]. And *Hieron. Welſchius*, one of a later date, of which himſelf was an eye-witneſs. "On the 16th of *December* "1631, when a very great earthquake was felt, "and terrible thunderings were heard at *Naples*, "a little before the next day-break *Veſuvius* was "ſeen to blaze out, being burſt open in ſeveral "places, notwithſtanding the thunder and earth-"quake ſtill continued[k]." But beſides *Ætna*, now *Monte Gibello*, and *Veſuvius*, or *Veſeuvus*, now *Monte* or *Montagna di Somma*, *Hecla* in *Iſland*, and others, ſeveral more ignivomous mountains or volcano's have been diſcovered within a few centuries. The *Sulfero* hill, or rather the field fum-

[f] *Senec.* lib. vi. nat. quæſt. cap. 4. [g] Loc. citat. [h] *Varen.* ubi ſupr. [i] Meteor. lib. ii. t. 42. [k] Itiner. ſui. p. 80.

ing

ing and burning with ſulphur near *Puzzoli*, called the *Solfatara*, as likewiſe *Stromboli* or *Strongylus*, according to *Welſchius*[l], was quite burnt out, fallen flat, and covered with the ſea about 30 years ago, before which it was ſurrounded with 8 other ſulphury hills (by the ancients called *Inſulæ Æoliæ*, and *Vulcaniæ* and *Lipareæ*,) one of which the ſame *Welſchius* ſaw burning, together with *Strongylus*[m]. Several have been found in the iſlands of the *Eaſt Indies*. One for example in *Java* burſt out in the year 1586, with a violent eruption of burning ſulphur. Mount *Gonnapi* in one of the *Bandan* iſlands, after it had continued burning ſeventeen years, was then rent aſunder, with an impetuous diſcharge of ſtones and ſulphureous matter. In the *Molucca* iſlands are many volcano's, the chief of which is the *Caminus Ternatenſis* before ſpoken of: All of which *Varenius* recounts at large from *Maffei*; and adds, that one of prince *Maurice*'s iſlands, near the *Molucca's*, is frequently viſited with earthquakes and eruptions of fire and aſhes. The like ſort of volcano's alſo abound in *Japan* and its neighbouring iſles, and in the *Philippines*; but moſt of all in *America*; nor have they been wanting, tho' at this time extinguiſhed, in the *Flanderkin* iſlands[n].

Phæn. VIII. Sometimes vaſt torrents of water flow out at theſe ruptures, forming lakes and ri-

[l] Itiner. p. 104. [m] Itiner. p. 195. [n] See *Varen.* geog. lib. I. cap. 10. prop. 5. *Athan. Kircher*. Mund. ſubterran. lib. ii. cap. 11. lib. iv. ſect. i. cap. 5 and 7. and præf. cap. 3. alſo *Bern. Cæſius* lib. i. de mineral. cap. 8. ſect. 2.

vers

vers where there were none before; and drowning whole cities and islands, which is confirmed by *Seneca* [o]. And *Aristotle* affirms, "that waters have "burst forth from the ground at the time of "earthquakes [p]". And the treatise *de Mundo* says, "Some earthquakes have opened foun-"tains where there were none before [q]". For examples of this kind read *Kircher* on the stinking lake which covers the city of *Euphemia* [r], and *Gassendus*, and *Furnerius* on the *Peruvian* earthquake, as above cited. Of the overwhelming of *Bura* and *Helice* in the *Corinthian* gulph *Pliny* makes mention [s], as also *Seneca* [t] after *Calisthenes*. Concerning the deluging the island *Atalanta*, see also *Seneca* from the account of *Thucydides* [u]. And *Plato*'s *Timæus*, and *Kircher* [w] of the *Atlantis* overwhelmed in like manner by an earthquake. They were such phænomena's as these, that possess'd *Democritus* and the ancient poets with the notion, that the subterranean waters were the original cause of earthquakes, and made them give *Neptune* the appellation of ἐννοσίγαιον καὶ σεισίχθονα, the *mover and shaker of the earth*, according to *A. Gellius* [x].

Phæn. IX. Winds and flatus's have also been observed to forego or accompany ruptures of the earth.

In earthquakes, says lord *Verulam* [y], "A cer-

[o] Lib. citat. 4. [p] Ἤδη καὶ ὕδατα ἀνεῤῥάγη γιγομένων σεισμῶν. Meteor. ii. t. 48. [q] οἱ δὲ πηγὰς φαίνουσι πρότερον οὐκ οὔσας. [r] Tom. i. pag. 77. tom. ii. pag. 257. [s] Lib. i. cap. 92. [t] Cap. 23. [u] Cap. 24. [w] Lib. ii. Mund. subterran. cap. 12. [x] Noct. Attic. lib. ii. cap. 28. [y] Hist. of winds.

"tain

" tain unuſual and unwholeſome wind has been
" obſerved before the eruption, as a ſweltering
" ſmoak breaks out before, and remains after
" great fires.". And *Seneca* ſays [z], " that often-
" times, when earthquakes are attended with any
" opening, wind will iſſue for many days, which
" thing is ſaid to have happened in the earthquake
" of *Chalcis*, as may be ſeen in *Aſclepiodorus*, who
" ſtudied natural philoſophy under *Poſſidonius:*
" And other writers will inform you, that when
" an aperture has been made in the earth, wind
" has iſſued out of it ſoon after, or, in other words,
" it eſcaped by a paſſage which it procured itſelf."
Of this examples have been given above, and *Seneca* himſelf ſays , " that there was ſomething
" of a venomous nature in the blaſts which ac-
" companied the earthquake in *Campania*, (which
" was the occaſion of his writing his ſixth book
" of *Natural Queſtions*) whereby a flock of 600
" ſheep was deſtroyed in the *Pompeiana Regio.*"

Phæn. X. On the other hand rivers, fountains and lakes have vaniſhed away from the places they formerly poſſeſs'd; ſeas have deſerted their wonted ſhores, at leaſt for a ſeaſon; and new iſlands have emerged where the waters uſually flowed without interruption. I call *Seneca* for a witneſs [b], who aſſerts that in his own days the iſland of *Theraſia* aroſe out of the *Ægæan* ſea, in the ſight of ſeveral mariners [c]. To which may not improperly be referred the origin of *Sicily* on the *Italian*, *Eubœa*

[z] Lib. vi. cap. 17.
[a] Cap. 1.
[b] Lib. citat. cap. 4.
[c] Cap. 21.

on the *Bæotian*, and *Cyprus* on the *Syrian* coaſt, of which *Pliny* [d], after he had proved the preſent poſition in a preceding chapter. Of the diſappearing of rivers and lakes in modern times, we have already mention'd a notable inſtance in *Peru*, from *Gaſſendus* and *Furnerius:* And there is a ſignal and a recent example of new iſlands, formed about the beginning of *July* 1686, as may be ſeen in *Gaſſendus* [e]. Thus the volcano of *Sicily* has produced a kind of offspring, or new little mountain, thence called *Volcanello*, as we learn from *Kircher* [f]. And the ſame hiſtorians relate that the ocean receded and returned with a great ſwell ſoon again, before the often mentioned earthquake in *Peru*; and further, that the ſame thing happened in the port of *Naples* before the raging of *Veſuvius* in 1631; inſomuch that *Hieronymus Welſchius*, a ſpectator of this uncommon ſcene, ſays, " that " ſeveral ſhips were in great danger of periſhing, " by being ſuddenly let down on land by the " retreat of the ſea [g].

Phæn. XI. Sometimes the duration of earthquakes is exceeding ſhort, conſiſting of no more than a few pulſes. Some again have laſted whole days, and even months and years, by fits. " If " they are not ſoon over, ſays *Pliny* [h], they may " probably laſt 40 days, and even longer, for " ſome have not wholly ceas'd in leſs than one, " and ſometimes two years; and this he repeats

[d] Lib. ii. cap. 88.

[e] In x. Laert. p. 1051.

[f] Loco citat.

[g] Itiner. p. 81.

[h] Lib. ii. cap. 82.

" in

" in another place[i]." *Ariſtotle* ſpeaking of the more violent ſort[k], maintains, with *Pliny*, that they do endure about that ſpace[l]. Notwithſtanding, this is what rarely happens; and although the earthquake of *Campania*, whereof he writes[m], did indeed continue ſeveral days, yet it does not appear to have held out altogether ſo long, nor did that other which overſpread *Sicily* in 1537, exceed 11 days; and laſtly, that which *Gaſſendus* obſerved at *Aix* in 1617, the night following the 13th of *January*, was quite over in leſs than three quarters of a minute.

Phæn. XII. They do not attack one ſingle place, but for the moſt part extend themſelves to ſeveral cities and countries very diſtant from one another, tho' they exert various degrees of violence at the very ſame time; and this was abundantly confirmed in our late inſtance. For all accounts agree that it was firſt felt at the very ſame inſtant of time, at *Lindau*, *Kempten*, and many other places, as at the cities and towns abovementioned; but in how different a manner it diſplay'd itſelf according to their ſeveral diſtances from *Hall*, where the ſcene was moſt dreadful, may be collected from the beginning of this diſcourſe. The ſame was obſervable in that of *Campania*, which *Seneca* deſcribes[n]. " *Pompeij*, " a conſiderable city of *Campania*, ſays he, was " thrown down by an earthquake, and the ſhock

[i] Lib. ii. meteor. [k] *ὅταν δισχυρὸς γένηται σεισμὸς*, &c. text. 45. [l] *μέχρι περὶ τεσσεράκοντα ἡμερῶν.* [m] Cap. 30. [n] Lib. vi. quæſt. nat. cap. 1.

" was perceived at the ſame time through all
" the adjacent country: And a little after part
" of the town of *Herculaneum* fell, and what con-
" tinued ſtanding, remains in a tottering condi-
" tion; and notwithſtanding none of the inhabi-
" tants of *Nuceria* loſt their lives, yet their miſ-
" fortunes were to be pitied: *Naples* had but a
" ſmall ſhare in the diſaſter, and the villages ele-
" vated on the adjacent hills, were ſenſible of the
" ſtroke, without any damage at all." In another place [o] he ſays, " when *Chalcis* was ſhaken,
" *Thebes* continued unmoved; *Ægium* reel'd two
" and fro, at the ſame time that *Patræ*, its near
" neighbour, felt not the leaſt motion, *&c.*" and
concludes, " that ſuch motion never is extended
" to the diſtance of 200 miles." Which if it always
held true in thoſe days, it no longer does ſo now:
For *Gaſſendus* takes notice, in the place above cited, that " not far from *Lima* (which, if I rightly
" remember, had then lately ſuffered an almoſt to-
" tal ſubverſion) there happened an earthquake
" which ran 300 leagues along the coaſt, and
" more than 70 into the continent," to which
add ſome other particulars which will be found
under *Obſ.* I. cited from *Meterranus* and *Kircher*.

Phæn. XIII. Mountainous places near the ſea
are chiefly expos'd to the moſt violent earthquakes;
whilſt flat, marſhy, inland countries, ſeldom or
never feel any ſhocks, at leaſt no original ones.
The ancients, as *Ariſtotle*, *Pliny*, &c. looked upon
Ægypt, *Gaul*, the iſle of *Delos*, &c. as quite ex-
empt from ſuch viſitations: Yet *Seneca* [p] aſſerts

[o] Cap. 25. [p] Cap. 26.

on

on the contrary, and experience proves earthquakes happened in all theſe places, tho' ſeldom, and in a milder degree. At *Alexandria* near the *Nile* in *Ægypt*, for example, about the year 551, and near *Bourdeaux* in *France*, in 584, according to *Garcæus*[q]. Nay we read in *Kircher*[r] that in the year 1660 in the month of *June*, an earthquake was propagated from this laſt city as far as *Narbonne*. What we have advanced concerning maritime and mountainous places, is confirmed by *Ariſtotle* in ſeveral examples[ſ] to which *Pliny* aſſents[t], remarking, that "though ſea coaſts are "obnoxious to the ſevereſt ſhocks, yet are not "mountainous ſituations altogether free from "them;" which he proves from the *Apennine* mountain and the *Alps*, which latter were not long ſince the theatre of ſuch like devaſtation. And *Seneca* alledges *Pompei* and *Herculaneum*, *Paphos* and *Cyprus*, *Tyre* and *Sidon*, as other examples[u]. *Peru*, *Campania*, *Calabria*, *Sicily*, &c. have been mentioned above as maritime countries, and abounding in mountains. As to marſhes, muddy and ſandy countries, as *Egypt* and *Tuſcany*, *Kircher* may be conſulted[w]. And the country about *Nurenberg* may teſtify for itſelf. As for *Garcæus* his obſervation, that the more ſouthern parts of the world are leſs obnoxious to earthquakes, than the northern, he is much in the wrong, as may

[q] Meteor. p. 389 and 405. [r] Mund. ſubterran. 257.
[ſ] Lib. ii. meteor t. 42. περὶ τούτους τοιούτους οἱ ἰσχυρότατοι γίγονται τῶν σείσμων, ὅπου ἡ θάλασσα ῥοώδης, ἡ ἡ χώρα σομφὴ καὶ ὑπαυλέος [t] Lib. ii. cap. 80. [u] Cap. 26.
[w] Mund. ſubterr. tom. i. p. 222.

appear not only from ſeveral of the foregoing remarks, eſpecially in *Phæn.* VII. but even from his own catalogue [x].

Phæn. XIV. It is furthermore certain that earthquakes have happen'd at all ſeaſons of the year, by night and by day, and under all varieties of conſtellations, indifferently. *Ariſtotle* [y], and *Pliny* [z], who in this matter almoſt copies him, are of opinion that moſt of them fall out in ſpring and autumn, oftner in the night than in the day-time, eſpecially a little before day-break. Our example, it muſt be allowed, confirms the latter, but then it ſeems to contradict the former; it attacking us in *July* in the very heat of ſummer, in the morning twilight, one hour after the change of the moon, no other remarkable aſpect offering at that time, except an approaching conjunction of *Jupiter* and the Sun, which the aſtrologers reckon no malevolent one. *Kircher* has theſe notable paſſages on this ſubject [a]. "As for what "*Ariſtotle* advances as to the time of earthquakes "happening, of iſlands in the middle of the ſea "being at all times without them, and their laſting 400 days, as it is contradictory to experience, we muſt not altogether rely upon it: For "they are not only places near the ſea, and iſlands juſt disjoin'd from the continent, that are "viſited by earthquakes; but they happen in the "very heart of large inland countries, and at all "ſeaſons of the year; a thing ſo well ſettled from

[x] Meteor. p. 393 & ſeqq. [y] Lib. ii. meteor. t. 41. [z] Lib. ii. cap. 80. [a] Tom. i. lib. iv. Mund. ſubterran. ſub finem cap. 10 ſect. 1.

"ob-

" obſervation and experience, that it admits of " no manner of doubt." *Seneca*'s words are very expreſs, " that the city of *Pompei* fell by an " earthquake in the winter, (to wit on the nones " of *February*) tho' our fore-fathers pronounced " that ſeaſon to be void of any ſuch danger[b]." *Tacitus* aſſures us that the earthquake which threw down the twelve cities of *Aſia*, came in the night; on the contrary that which *Kircher* himſelf ſaw, was in the day-time. That at *Lima* in *Peru* was in the winter on the 24th of *November*, five days after the new moon, *Mars* and the moon being in conjunction, but the moon at the ſame time in quartile to *Mars*, and in ſextile to *Mercury*. Now let any one who has leiſure, conſider well the ſeveral examples adduced above; after which let him carefully peruſe *Garcæus*'s catalogue of earthquakes, each accompanied with its concomitant configuration of the heavens from *Ephimerides*; and I am ſatisfy'd he will be convinc'd, that there is no ſeaſon of the year, nor any celeſtial configuration under which an earthquake may not happen, as well as at any other time.

Phæn. XV. After a very ſevere earthquake has happened, attended with a great conflagration, ſuch another does not ſucceed in a ſhort ſpace of time, but for the generality after a long interval, and then eſpecially when a neighbouring volcano that was almoſt extinct, flames out afreſh, or affords tokens of doing ſo ſoon. *Seneca*[c] proves the truth of this obſervation, and *Campania* and *Sicily*, *Ætna* and *Veſuvius* are vouchers, as well as ſeveral other

[b] Cap. 1. [c] Lib. citat. cap. 31.

places mentioned in *Phæn.* VII. See likewiſe the writers there cited. It is remarkable, by the bye, that ſeveral volcano's which formerly threw out fire, are now utterly extinguiſhed. The iſland of *Querimodam* on the *Braſil* ſhore, not far from the river *Plata*, for example, as alſo certain mountains in *Congo* and *Angola*. Geographers reckon ſeveral places among the *Azores*, eſpecially in *Tercera* and St. *Michael*, which formerly flamed out, but of latter days have emitted nothing but ſmoak, which alſo has ceaſed in ſome of them; whence we may infer that ſome parts of the earth may in time get rid of ſuch accidents; *Ariſtotle* [d], I know, thinks the thing impoſſible, but I can perceive no reaſon why he ſhould do ſo.

Phæn. XVI. It is ſaid that fiery meteors have been the forerunners, and ſometimes the concomitants of earthquakes: Alſo a continually clouded ſun, a turbid foulneſs of wells and fountains, infected with a filthy ſaline taſte, a deſertion of animals and birds, *&c.* and that to theſe have ſucceeded, peſtilences, contagious diſeaſes, famine, ſedition, and a train of other evils: Of which *Pliny* [e], *Ariſtotle* [f], *Seneca* [g], *Garcæus* [h], and others. Nothwithſtanding which it would be well worth our inquiry, to examine well if theſe things have really at all times or for the moſt part, any natural connexion one with another, or that it was by mere accident that they preceded or followed after. I ſhall ſet down ſome modern inſtances.

[d] Lib. meteor. t. 40. [e] Lib. ii. cap. 81. [f] Lib. ii. meteor t. 42, 43. [g] Loc. cit. [h] In catal. terræmot.

When

When *Vesuvius* raged in 1631, *Welschius*[i], who was present, observed that the sun was darkened, and a general duskiness was diffus'd through the whole atmosphere, from the very copious eructation of ashes; so that it seemed to look as if lightnings were glanced from a cloud which covered the head of the mountain; and it was confirm'd by abundance of letters from *Italy*, that the same scene was repeated again, tho' with a much more horrid appearance, in the month of *July* 1660. Thus the mountain in the island of *Java*, which in the year 1586 was riven asunder by a violent eruption of burning sulphur, sent forth such a vast quantity of thick black smoak for three days together, mixt with flame and fiery sparks, as obliterated the sun, and almost turned the day into night[k]; and the like was observed on another mountain called *Gonnapi*. To which may be here added the relation sent from *Ternate* to *Bandam*, of the rending of mount *Gammacnorra*, as before recited in *Phæn.* V. So also in those most horrible earthquakes which afflicted the inhabitants of *Santorini* in the *Archipelago* in 1650, from the 24th of *September* to the 9th of *October*, the sky was darkened, and the air infected with stinking sulphureous vapours, to such a degree, as blinded every body that ventur'd out of doors, for three days together[l].

[i] Itiner. p. 80. [k] *Varen.* lib. x. cap. 10. prop. 5.
[l] Test. *P. Francisc. Riccardo*, in mund. subt. *Kircher*. p. 182.

HYPOTHESIS

Framed for ſolving the foregoing PHÆNOMENA.

I Shall ſay nothing of *Bodinus*'s dream of evil genij, mention'd by *Honoratus Faber* [m], nor of the *Japoneſe* dragon ſo largely treated of in the embaſſy to that iſland, nor of *Thales*'s conceit of the earth floating in water as a ſhip on the ocean, and of the toſſings ſhe now and then undergoes. But I muſt obſerve that *Democritus* of old, and ſome others, whoſe doctrine as to this matter were not much oppoſed by *Epicurus*, and in a manner aſſented to by *Seneca* [n], held that there are mighty rivers continually running, and vaſt oceans in a perpetual agitation below in the ſubterranean regions, and that, when any colluctation happens there, the earth of conſequence muſt tremble and ſhake; and that the ancient poets had this notion, is evident from *Aulus Gellius* [o]. *Anaximenes* affirm'd that "the earth was the cauſe of its own "motions, by letting ſome of its parts drop into "its cavities, which were either diſſolv'd by wa-"ter or prey'd upon by fire, or driven about by "winds, or deſtroyed by time [p]." Others maintained, with *Archilaus*, that winds inſinuating themſelves into the bowels of the earth, do there impel the compreſſed air, and force it to break through its confinement.

[m] Tract. vi. prop. 22. [n] Lib. vi. nat. quæſt. cap. 7, 8.
[o] Lib. ii. cap. 28. [p] *Senec.* cap. x. c. 1.

II. A

II. A like opinion prevail'd in the *Peripatetic* ſchool for ſeveral centuries. And *Seneca* himſelf did not deny the ingreſs of winds from without, although he aſcribed theſe calamitous accidents rather to ſubterraneous exhalations and vapours [q]. For the notion ran, that there was a conſtant evaporation from the earth, ſometimes dry and ſometimes combined with moiſture. When this was ſent up from below, and raiſed as far as it could go, and meeting with an obſtruction, was forced back upon itſelf, then conflicts and tumultuous motions aroſe. To this point likewiſe tended *Ariſtotle*'s hypotheſis, as appears plainly in his metereologies [r]. For he ſets out with aſſerting, that both moiſt and dry exhalations are raiſed within and about the earth, and when theſe are over copious they produce earthquakes. For the earth being ſaturated with moiſture, and heated by the ſun without, and by fire within, πολύ μεν ἔξω πολὺ δ' ἐνῖος γίνεσθαι τὸ πνεῦμα. Καὶ τȣ̃το ὁτὲ μὲν συνεχὲς ἔξω ῥεῖν πᾶν. ὁ τὲ δὲ εἴσω πᾶν. ἐνίοῖε μερίζεσθαι. That is, *much ſpirit is generated without, and much within. Sometimes this is diſcharged entirely outwards, ſometimes it is abſorbed inwards, and ſometimes it is divided.* Which, as he ſeems to have advanc'd for want of ſomething better, he endeavours to puzzle the cauſe. Now, we are to conſider, ſays he, ὁποῖον κινηῖικώτατον ἂν εἰν τῶν σωμάων; *what is that body of all others that is moſt ſtrongly diſpos'd to motion? Why doubtleſs*, he anſwers, τὸ σφοδρότατον, *that which is moſt violent*, and ſuch he concludes to be τὸ τάχιςα φερόμενον, *that which moves ſwift-*

[q] Cap. 13 and 23. [r] Lib. vii. cap. 8.

eſt, and τολεπ]ότα]ον, *the moſt ſubtile and penetrating*: ὥςε ἔιπερ ἡ τȣ̃ πνεύμα]ος φύσις τοιάυ]η, μάλιςα τῶν σομάτων: *Since then this is the moſt apteſt of all bodies to motion.* From whence he deduces this final concluſion, ȣ̓κ ἄν ȣ̃ν ὕδωρ, ȣ̓δὲ γῆ αἴτιον εἴη, ἀλλὰ πνεῦμα, τῆς κινήσεος, ὅταν ἔσω τύχῃ ῥυὲν τὸ ἔξω ἀναθυμιώμενον. *Wherefore neither water nor earth can be the cauſe of* (its own) *motion, but ſpirit*, (or vapour) *when, by any accident, the external exhalation is turned inwards.*

III. The greateſt defect of *Ariſtotle*'s hypotheſis, is that he unluckily never thought of an actual accenſion, or kindling of the dry exhalations excited within the earth, which the inflammation of gunpowder might have hinted to him, had he been acquainted with it: Yet he could not but have been well informed of the burning of *Ætna* and *Lipara*; and he moſt certainly was ſo, if the book περὶ θαυμασίων ἀκȣσάτων be his, which might have ſupplied him with the like notions as thoſe which occur in the book *de Mundo* [f]; unleſs, with *Heinſius*, we deny that *Ariſtotle* was the author of that treatiſe too, in which earthquakes are derived from ſubterraneous fountains of fire, much in the ſame manner as that whereby the modern philoſophers have endeavoured to account for them. Indeed the ancients according to *Seneca* [s], had. *Anaxoras* referred the cauſe of earthquakes to ſubterraneous clouds burſting out into lightnings which ſhook the vaults which confined them. Others, that the arches which had been weakened by continual fires, at length fell in, others de-

[f] Cap. 4. [s] Cap. 9.

riv'd

riv'd theſe accidents from the rarefied ſtream of waters heated by ſome neighbouring fires: and ſome, as *Epicurus* of old, (among the reſt of the opinions collected by *Seneca* [u]) and, as *Andreas Cæſalpinus* [w] ſhews, ſeveral of the Peripatetic ſchool alſo aſcribed theſe horrible accidents to the ignition of certain inflammable exhalations.

IV. And this has been the favourite hypotheſis of the moſt celebrated modern philoſophers, *Gaſſendus*, *Kircher*, *Schottus*, *Varenius*, *Des Cartes*, *Du Hamel*,, *Honoratus Faber*, &c. Though it ſhould be noted, that this laſt imagines that waters extremely rarefied by heat, may ſometimes force a way through their proper boundaries, and that included vapours may, under the like circumſtances act in the ſame manner, and ſo be ſometimes alſo productive of earthquakes. Theſe learned men do ſuppoſe that there are many vaſt cavities under ground which have a communication with one another by intermediate canals, ſome of which abound with waters, others with vapours and exhalations ariſing from inflammable ſubſtances, as bitumen, nitre, ſulphur, *&c.* and alſo metals and minerals, congeſted together, at all times diſpoſed for inflammation, and on ſome occaſions in an actual ſtate of accenſion: All which doctrine is conſonant both to reaſon and experience, as will be preſently proved at large. Now whether ſuch combuſtible exhalations as theſe happen to be kindled up by any ſubterraneous ſpark, or from ſome active flame gliding thro' a narrow fiſſure from without, or in conſequence of the fer-

[u] Cap. 20. [w] Lib. iii. quæſt. peripatet 9.

mentation

mentation of ſome mixture, they muſt neceſſarily produce pulſes, tremors, or ruptures at the ſurface, according to the number and diverſity of the cavities, and the quantity and activity, *&c.* of the inflammable matter: *Honoratus Faber* illuſtrates this doctrine by a variety of artificial earthquakes, as he calls them, confining gunpowder, (a mixture of nitre, ſulphur and charcoal) in pits, and ſetting fire to it by a train [x].

The laſt mentioned hypotheſis I acknowledge for my favorite; being ſuch as the nature of burning mountains, as well as of thoſe parts of the earth, moſt liable to earthquakes do plainly indicate; for they all abound in ſulphur, nitre, bitumen, and the like inflammable ſubſtances. This, of all the reſt, has the advantage of ſatisfactorily accounting for the ſeveral recited phænomena; to evince which I ſhall premiſe a few obſervations, as principles of future concluſions.

I. The earth incloſes great numbers of ſpacious cavities, vaults and canals, eſpecially under the ſummits of mountains. To paſs by the famous *Specus Coricianus* ſpoken of by *Mela*, *Solinus*, *Pliny*, *Strabo*, &c. *Pluto's Den*, mentioned by *Ælian*, and other ſubterraneous hollows ſcattered up and down in *Seneca*, the amazingly extended caverns under the *Andes* in *America*, and thoſe in ſome parts of *China*, deſcribed by *Martinius* [y], and more at large by *Kircher* [z], I prove my aſſertion from thoſe ſtrange ſpiracles, called, from the continual blaſts

[x] Lib. citat. Prop xxx. Vide etiam Gaſſend. Phyſic. Sect. III. Memb. I. Lib I. Cap. 6. p. 48, 49. [y] In Ablante Sinic. [z] Mund. Subterr. Lib. II. ſub finem.

 they

they ſend forth, the *Æolian Bellows*, which the ſame *Kircher*[a] both deſcribes and ſaw; I prove it from the innumerable ſources and waters every where abounding; and laſtly, I prove it from the vulcano's and burning mountains diſtributed through many regions of the world, as *Italy*, *Aſia*, *Media*, *Tartary*, *Japan*, the *Philippines*, and other parts of *India*, *Africa*, *Terra Auſtralis*, *Mare del Zur*, the *Canaries*, *North* and *South America*, *Greenland*, *Iſland*, &c. of which according to authors of the beſt credit cited in Phæn. VII. there is an immenſe number: And one thing is to be particularly remarked, that the cavities of theſe burning mountains do not terminate at their baſes, but are far extended in canals which often communicate with one another. When mount *Ætna* of old begun to emit flames, *Strongylus* in the *Liparæ* Iſlands did the like at the ſame time, the ſulphureous ſteams diffuſed under all *Sicily* taking fire at once: And altogether as remarkable, or more ſo, is *Kircher*'s obſervation, concerning that moſt terrible earthquake in *Calabria*, which himſelf ſaw and felt, that *Strombulo*, 60 *Italian* miles diſtant, was not only heard to bellow and ſeen to blaze a little before, but that the ſubterraneous noiſe was firſt diſtinguiſhed but dully, and then waxed louder and louder, till it arrived under the ſpot on which he and his companions ſtood.

Obſerv. II. Some of theſe caverns and ſubterraneous paſſages, when replete with water, form gulphs, abyſſes and rivers, and ſome give riſe to ſprings; others are occupied by flatuſes and exha-

[a] Loc. cit.

lations;

lations; and others again with fire and flames, as hinted in the proofs of the foregoing obſervation. But for further confirmation, of what relates to waters, it will demand but a moderate degree of ſagacity to conceive what vaſt reſervoirs of that fluid lie under the *Alps* for example, which pour forth ſo many great rivers, as the *Danube*, the *Rhine*, the *Inne*, the *Rhone*, the *Saone*, the *Maeſe*, the *Moſelle*, the *Po*, the *Etſch*, the *Mencio*, the *Teſino*, the *Save*, the *Drave*, &c. beſides the great lakes of *Swiſſerland*, as the *Lucern*, the *Lemann*, that of *Zurich*, and the leſſer ones as you enter *Italy*. The concavities under *Taurus*, *Antitaurus*, *Caucaſus*, and *Imaus* in *Aſia*; whence flow the *Indus*, *Ganges*, *Oxus*, *Hydaſpes*, and ſeveral rivers of *China*; likewiſe the *Euphrates*, *Tigris*, &c. The like under the *Mountains of the Moon* in *Africa*; whence the *Nile*, the moſt celebrated of all rivers; the lakes *Zaire*, *Zembre*, &c.——Under the *Andes* in *America*, which pour out a profuſion of mighty rivers and lakes on every ſide, herein exceeding all others. Whence it is eaſy to imagine what an infinity of other leſſer receptacles of water there muſt exiſt throughout this globe, whence rivers of leſſer note are derived; and, if the earth be properly called the *Terraqueous Globe*, ought it not to be ſo? As to the ſecond part, which concerns flatuſes and exhalations, peruſe what *Gaſpar Schottus*, a diſciple of *Kircher*, writes about artificial winds generated *intra Æolias Cameras*, by the fall of water [b], and then judge what quantity of winds muſt of neceſſity be continually excited in the bowels of the earth

[b] *Magia Muſica*. Syntag. IV. cap. 10.

 from

from the boisterous dashing of the ocean against the shores, and the ingress of its waves into the subterraneous caverns. As to the third part, fire, consider, 1. what a vast plenty of hot springs is there in all parts of *Germany*, *France*, and *Spain*. Can these receive their heat, as it were by accident, merely from the abysses of volcano's prolonged through an innumerable variety of canals, or must not they owe it to a more extensive infernal fire? 2. *Æstuaries* and eruptions of fire are to be seen at *Petra mala*, and about *Puzzoli* in *Italy*, and in many other places, and sometimes they have been known to be thrown up from the bottom of the sea, as in 1650, and long before in 1457 and 1570, near the island of *Santorini* in the *Archipelago*. 3. It should be observed, that all these things are agreeable to the oeconomy of nature. The whole stock of waters under the surface of the earth would be converted into ice, if some of them were not exceedingly heated by the proximity of subterraneous fire. Again, These very fires would be extinguished, were it not for the recreating blasts of air, produced by the ocean as before hinted, or admitted in through the apertures of volcano's. And finally, there would be a total consumption of all, from the same fires, were they not restrained and partly extinguished, by the intervention of waters and humid vapours. I might here recite a notable passage to this purpose out of the book *de Mundo* [c], and another from *Andreas Cæsalpinus* [d], had I not so long insisted in the proof of this 2d observation.

[c] Cap. 4. [d] Lib. III. Perip. Quæst. IX. p. 77.

Observ.

Obſerv. III. The bowels of the earth do every where, but chiefly in mountainous places, hold more or leſs of ſulphur, bitumen, nitre and other ſalts, amber of various kinds, &c. alſo divers metals, and that in great plenty; but theſe ſubſtances are obſerved to abound moſt of all in countries which have been viſited with the ſevereſt earthquakes. Natural geography and experience teach us, that all *Sicily*, *Campania*, *Tuſcany*, and indeed *Italy* in general, have plenty of ſulphur, bitumen, coals, pumice ſtones, iron, copper, and other ores, and the like holds good of many others. Whereſoever burning mountains are found, and we have ſeen above that few parts are without them, there theſe inflammable minerals are even belched forth: And it is very remarkable that the Iſle of *Ormus* in the *Perſian* Gulph, which geographers report to be in a manner all ſalt, did not only burn ſeven whole years together, but does even to this time daily throw forth balls of flame from its ſaline mountains, a certain token of the truth of that obſervation among naturaliſts, that foſſile ſalt is rarely found pure, and void of all metalline mixture, or a degree of unctuous fatneſs. Nor need I mention that *Pliny* and *Albertus Magnus* affirm, that oil may be extracted from ſalt, and ſalt from all metals and earths; or alledge a curious and a deciſive experiment to prove that the earth every where abounds with fatneſs and the *pabulum* of flame [e]. It were needleſs here to ſay any thing of the mines and minerals of *Germany* and its neighbouring countries, of which the geographic writers

[e] Vide Kircher, Mund. ſubterran. Tom. I. p. 185.

are

are ſo full. I muſt however take notice, that in *Miſnia* there is a mountain of coals, which frequently ſends forth ſmoke, and ſometimes actual fire, whoſe flames about the year 1505, *Agricola* the great mineraliſt ſaw raging to an exceſſive height. And *Bernh. Cæſius* writes [f], that frequently in the night ſeaſon flames break out and blaze through the whole tract of land between *Zwiccaw* and *Glauch.* Which writer gives a very large account of the ſeveral countries of the world. that principally abound in ſulphur, bitumen, ſalt, ſuccinum, and other minerals and metals. This one thing more I have to add, that from the fiery eruptions at *Santorini*, ſpoken of above, it is manifeſt that even ſubmarine places are not entirely deſtitute of ſulphureous and bituminous minerals: And that the frequent appearance of fiery meteors, in every part of the known world, afford a general argument for the exiſtence of ſuch inflammable ſubſtances every where under ground; for all naturaliſts allow that they can be no other than ignitions of ſuch exhalations.

Obſerv. IV. It ſtands therefore with reaſon as well as experience, that the ſubterraneous cavities and paſſages are full of exhalations continually and copiouſly raiſed from theſe inflammable bodies, and that ſuch ſteams are no leſs inflammable than the bodies themſelves which they are produced from, whether they happen to be kindled by ſome fortuitous ſubterranean ſpark, or from the fermentation of the ſteams of different bodies: For as they are

[f] Lib. I. de Minera. cap. 7.

elevated

elevated as high as the middle region of the air, where they can meet with no fire to ignite them; what is more probable than that this operation is performed in the under regions of the earth? Moreover, that vaſt quantities of the exhalations of ſuch bodies are congeſted in the bowels of the earth is evident from this alone, that ſulphur can never be dug deep under ground, but only from mines expoſed to the open air and day-light, otherwiſe the miners would be ſuffocated thereby; and on the ſame account all places in the neighbourhood of the *Aſphaltites* lake are abſolutely uninhabitable. That an ordinary candle is capable of ſetting ſuch ſteams in a blaze, is obvious in *Naptha*, a few drops of which as ſoon as poured out, will ſpread alſo a pinguous vapour through whole ſtreets, producing an inflammation in the air wherever it reaches. And laſtly, that ignition may ariſe out of mere fermentations, without the preſence of any actual flame, is proved from the eaſily kindling up of a mixture of nitre, ſulphur and quick lime, by moiſtning it with a little water or ſpittle [g]. It is further very remarkable, that not only ſeveral of theſe inflammable ſubſtances either by themſelves or mixed with others, will burn in the midſt of water; but that even gold, and other metals, minerals, *&c.* duly prepared, will be eaſily put in a ſtate of accenſion not only by fire, but by a moderate degree of warmth alone, and thereby produce amazing effects; ſuch as I have myſelf

[g] Vide *Gaſp. Schott.* mag. pyrotechn. p. 121.

more

more than once beheld, and of which the aforecited author treats at large[h].

Obſ. V. The force of ſuch inflammable vapours, to produce motion, and alſo pulſations and ſhocks, when in a ſtate of actual accenſion, is prodigious. " The power of gunpowder fired in ordnance or in " mines, is well known: That it is capable of over-" ſetting and blowing up the moſt ſolid founda-" tions. And if we examine into the cauſe of ſo " vaſt an impulſive force, we ſhall find it to reſide " in nothing but a compoſition of a little nitre, " ſulphur and charcoal. But if there be ſo much " ſtrength in a ſmall quantity of this artificial " powder, how immenſely greater may we not " ſuppoſe that to be, which ariſes out of nature's " treaſure of combuſtible materials of ſulphur, " nitre, alum, ſal ammoniac, bitumen, and other " ſpirits of minerals, metals, gold, copper, iron, " arſenic, quickſilver, *&c.* every one plentifully " ſtored up in the hidden cavities of the earth?" I uſe the learned *Kircher*'s words, as the apteſt to expreſs my meaning. Travellers who have viſited *Vulcan*'s fields near *Puzzoli*, give a horrible deſcription of the impetuous blaſts which ſome of thoſe ſpiracles belch out, with moſt aſtoniſhing noiſes, and with a force able to repel back into the air large ſtones thrown into them. What a huge crack do the fulminating powders of gold, copper, tartar, *&c.* produce in their exploſion; violently burſting to pieces whatſoever obſtacles they meet with? To ſay nothing of the dreadful and

[h] Mechan. Hydraulic p. 63. vide etiam *Gaſſend.* animadv. in Diog. Laert. p. 1016, *&c.*

penetrating energy of lightning; which the city of *Stralſund* in *Pomerania* not long ago ſadly experienced.

Obſerv. VI. The force of ſpirituous bodies in a ſtate of rarefaction, even without accenſion, is alſo very great: However, without the concurrence of ſome extrinſick impulſe, it ſeldom manifeſts itſelf in ſudden ſhocks and concuſſions; but chiefly in ſlighter tremors, ſometimes accompanied with ſimple ruptures of the ground. *Schottus* procur'd a ſort of little glaſs ſpheres to be made at *Rome*, and above forty years ago I diſtributed ſeveral of them among my friends at *Jena*, which I brought from *Amſterdam*. Theſe would give a report almoſt as loud as a muſquet. They were filled half full of vinegar or ſome ſpirit, and then hermetically ſealed. Being placed on burning coals or in hot embers, the liquor within, tho' rarefied by the heat, did not boil, or ſo much as move the ſphere, but, burſting its priſon at once, bounced as loud as a piſtol. Much in the ſame manner it comes to paſs that pillars of marble which the united force of an hundred yoke of oxen cannot pull aſunder, are by authors of good credit affirmed to be eaſily broken to pieces by the rarefaction of a little air or ſpirituous fluid lodged in their pores, when ſurrounded with fire; but at the ſame time they make not the leaſt mention of any tremors or reiterated pulſations preceding the diſruption.

Obſerv. VII. Metals and minerals are not only formed in the bowels of the earth, but after having

ing been removed, are again regenerated in the very ſame places. This is obvious to every day's experience, as may be proved from *Agricola*[i] and *Cæſius*[k]; eſpecially in the iſland of *Ilva* or *Elva* in the *Tyrrhene* ſea, where it has been obſerved that a mine entirely cleared of its iron ore, had it renewed in the ſpace of twenty five years: And lead gutters expoſed long to the open air on the tops of houſes, have been found to exceed conſiderably their original weight; alſo metalline ſhafts or adits wrought at firſt large enough to admit an eaſy paſſage to the miners, have in proceſs of time grown ſo narrow, as to be quite uſeleſs, which could no otherwiſe come to paſs, but by an acceſſion of new matter, according to the ſentiments of the now mentioned writers.

Obſerv. VIII. Mineral ſteams are indeed ſometimes found to be harmleſs, eſpecially when temper'd with an intermixture of bodies of a different nature: Yet for the moſt part they are obſerved to be noxious, eſpecially if over copious, both to men and beaſts. The former part of the obſervation is proved by the ſalubrity of hot ſprings and medicated waters, plentifully impregnated with ſteams of ſulphur, nitre, *&c.* Such are frequently met with in *Italy*; nor are they very ſcarce in *Germany* and other countries about it. The latter part is notorious from the number of diſeaſes which ariſe from metals and metalline fumes; ſome attacking the joints, others the lungs, ſome the eyes, and others again the whole habit, ſo as

[i] Lib. v. *de ortu cauſ.* ſubterr. & *conimbric.* tract. 13. met. cap. 2. [k] Lib. i. de miner. cap. v. ſect. 5.

to bring on death. By repeated obſervation it has been found, that in pits and quarries where ſtones have been broken by fire, the air is vitiated with a poiſonous infection, and the cracks and junctures of theſe ſtones do exhale a ſubtile virulent ſteam, which the fire forces out from interſperſed metalline particles, of ſuch a nature, that when any animal bodies are infected with it they ſwell, and loſe all ſenſe and motion. It is reported that near *Plana*, a town of *Bohemia*, there are grottos which at certain ſeaſons of the year emit a vapour which extinguiſhes lights, and kills the miners who tarry a ſmall time in it; and of the like nature is the ſoil about *Puzzoli*, and the famous *grotta di cani*, the lake *Avernus*, &c. ſeveral more of which are to be met with in the writings of *Bernard Cæſius* [l], *Athanaſius Kircher* [m], and *Seneca* [n].

Concluſion I. The earth being, (by *Obſerv.* I.) every where below hollowed out into caverns and canals, which (by *Obſ.* III, and IV.) includes vaſt ſtores of various metals, minerals, and readily inflammable ſubſtances; it may eaſily come to paſs, from the fire, likewiſe diffuſed through the whole bowels of the earth (*Obſ.* II.) that ſome little ſpark may from a great diſtance, by a chink or ſmall aperture, find its way into the ſaid caverns, and ſo ſet fire to the ſulphureous and nitrous ſteams, or that they may be kindled up by ſome ſudden fermentation: In either caſe it is evident (from *Obſ.* V.) that ſo ſudden an inflammation and rare-

[l] Lib. i. de miner. cap. vi. ſect. ii. [m] Mund. ſubterr. tom. i. lib. v. ſect. iii. [n] Lib. iv. quæſt. nat. cap. 18.

faction,

faction, muſt neceſſarily, according to the greater or leſſer quantity of combuſtible matters, their ſubſtance, tenacity, degree of drineſs, the extent, figure and poſition of the caverns, *&c.* produce various pulſations and other violent effects; repreſented, tho' in miniature, by gunpowder fired off in artificial mines, by a long train or match. And indeed in theſe days the knowledge of gunpowder has hinted the true cauſe of earthquakes in general, and of the various phænomena of particular ones, and that in a fuller and more ſatisfactory manner, than the ancients, for want of ſuch aſſiſtance, could any ways make out.

Concl. II. Nor is it ſtrange that ignes fatui, and other fiery meteors ſhould ſometimes be ſeen without any ſubterraneous accenſion, or enſuing earthquake; ſince the intervention of a little moiſture may eaſily ſtifle and extinguiſh ſuch ſudden inflammation; or ſuppoſing ſome ſubterraneous vapours to be actually kindled, their flames may find vent, and eſcape through ſuperficial crannies; juſt as the blowing up of artificial mines is frequently defeated by a dampneſs of the powder, or by a wrong proportion of the ingredients, or by the mine being too ſpacious for the quantity of the powder; or laſtly, if through the careleſſneſs of the engineers, or the craft of the enemy, there be any apertures whereby the flame of the kindled powder can find a vent.

Concl. III. But when ſpirits in a ſtate of actual inflammation are ſo confin'd as to have no paſſage at all to eſcape through, and at the ſame time the preſſure of the incumbent maſs, or the

 cohe-

cohesion of its parts be too great to yield to the impulse; the consequence then will be at least a commotion and tremulous concussion, in proportion to the said incumbent mass. And here, by the way, it may be observed, that since the caverns below the earth's surface, cannot in reason be supposed to bear any proportion to the whole globe; this alone may afford an easy solution of the second phænomenon.

Concl. IV. It is easily to be comprehended, that when the impulse is directed parallel to the horizon, or upwards perpendicular to the surface, or obliquely between both, it can force a passage through the obstacle no otherwise than from the various positions of the caverns and canals; that is, as they happen to point horizontally, vertically, or obliquely; just as in guns, the force of the powder is directed the same way that the piece is planted: And on this footing the diversities of general earthquakes mentioned at large in *Phæn.* III, IV, and V. will be satisfactorily accounted for.

Concl. V. Nor is it difficult to foresee, if it should so happen, as it very easily may, that a cavern transversly extended in length, should be ignited near its middle, so that the impetus must be directed at the same time to both its extremities, what would be the consequence; namely that those extremities receding farther asunder, must during the blast, produce a rupture in the roof above, which as soon as that was spent, would close again with a reciprocal force: And such is the cause of the *arietation* described in *Phæn.* IV.

Concl.

Concl. VI. It is likewife manifeft, that when any part of the earth fuffers fome degree of a fhock, or a confiderable trembling, even though the fuperficial part be not ruptured afunder, fuch fuperincumbent lofty ftructures as towers, churches, *&c.* muft be either thrown down or fhattered thereby: As when a table receives a fmart ftroke on the underfide, drinking glaffes placed thereon will be overfet; and nuts, fruit and the like, leap out of the plates that hold them. This fhews how the *fuccuffion* and *fubverfion* particularly defcribed in *Phæn.* III and IV. are to be rationally explained.

Concl. VII. But when the earthen roof is too weak to refift the efforts of a more furious accenfion, the flames muft needs burft open the gates of their confinement, and every thing upon the furface go pell mell to the bottom, the fides of the cavern at the fame time collapfing; and thus whole cities, mountains, rivers and even iflands, may be fwallowed, and all thofe horrible effects produced, which were enumerated in the five firft phænomena: Nor is the art of war practifed under ground, incapable of working fimilar confequences.

Concl. VIII. And further, fince it appears from *Obf.* II. that vaft refervoirs and torrents of water are contained in the fubterraneous apartments; what fhould hinder but that fuch a body of fluid may inftantly overflow the cities, mountains, *&c.* newly fwallowed up, and form large ftanding lakes, or flowing rivers, where there were no figns of them before? Which will fatisfy the latter part of of *Phæn.* IV. and alſo the whole of *Phæn.* V and VIII.

Concl. IX. But if an huge bulk of earth be forced up obliquely through the incumbent ſea, ſo as not to drop back into the ſubmarine cavern, but to reſt on the ſolid bottom near the aperture, with its top above the ſurface of the ſea, a new iſland will be formed; and if at the ſame time much of the ſea be abſorbed into the abyſs below, ſubmarine hills may have their tops uncovered, and thus alſo become ſuddenly new iſlands: And thus the cauſe of *Phæn.* X. may be naturally explain'd.

Concl. X. And to the very ſame cauſe muſt the ſea's inſtantaneous receding from the ſhore during an earthquake (as mentioned at the latter end of *Phæn.* X.) be aſcribed; it being ſucked into the new opened gulph below, and diſappearing 'till diſtant waves flow in and ſupply its place.

Concl. XI. Nor is it to be accounted ſtrange that when, and whereſoever earthquakes happen, flames ſhould not at the ſame time be always viſible: For theſe, if not extravagantly fierce and copious, may be ſmothered and extinguiſhed by the fallen ruins of the earth, or by the overflowing of waters: Beſides they may be often, either of ſo ſubtile a nature, or ſo involv'd in clouds of ſmoak, as in the day time to eſcape our ſight, though they might be viſible enough in the darkneſs of the night; of which *Ætna*, *Veſuvius*, and the fields of *Puzzoli*, do afford almoſt daily examples.

Concl. XII. Flames are a great deal more apt burſt forth from the tops of mountains, than in valleys or other low places, as being lefs check'd by the beforementioned obftacles, and likewife becaufe the cavities under mountains are very frequent and large, and their fides by inclining together, form a kind of chimneys which favour their afcent. This explains the former part of the VIIth *Phæn.*

Concl. XIII. And fince vaft quantities of fulphur, bitumen, ftones and metals, liquified by a moft intenfe heat, are expelled from thefe infernal chambers through the tops of mountains, like ftones and bullets out of artificial ordnance, they muft be the pabulum whereby fuch fire is fo long maintained, except that the crufty rubbifh which drops off from the inward lining of thofe mountains, may fometimes fupply it with new fewel. Hence the fecond part of *Phæn.* VII is deduc'd.

Concl. XIV. The caufe is likewife manifeft, why thefe ignivomous dragons, after having ceafed for a while, through a total confumption of the combuftible materials within them, do rage again: This being the confequence of another accenfion of newly generated fteams and exhalations, which, like the former ones, forces a new vent for other ignited and melted fubftances, as in Obf. VII. and thus the laft part of the fame VIIth *Phæn.* may be, at leaft probably, accounted for.

Concl. XV. With the like eafe may we conceive how ignited fumes and exhalations being rarefied in the bowels of the earth, do occafion tremors and fhocks, as alfo winds and blafts, fometimes before the ragings of burning mountains, and fometimes after them; namely, in the former cafe, case, because no vent is as yet opened; and in the latter, because it is closed up again before they have entirely escaped; and thus the force being distributed among the neighbouring parts, the inclosed air is driven out through whatsoever crannies it happens to meet with, as from æolipiles, and thus we have a very probable solution of the IXth *Phæn.*

Concl. XVI. Nor is it strange that such eruptions should be for the most part accompanied with horrible noises; we experience them in a proportionable degree upon discharging guns, exploding fulminating powders, and bursting bladders. And the variety of these noises, as bellowings, lowings, thunderings, roarings, *&c.* depend upon the different capacities and figures of the caverns and canals, like the various tones of an organ on the sizes and length of its pipes. Such is the cause of *Phæn.* VI.

Concl. XVII. Sometimes the chambers which contain the combustible matter are small and few, and their walls not so thin as to permit the kindled flame to make a sudden irruption into the contiguous ones, which rather burns a passage through by gradually consuming the intermixed sulphur and bitumen, and then perhaps meets with much more capacious caverns, through which being equally diffus'd, much of its primary force is abated, and its velocity retarded; which affords a satisfactory rationale of the different durations of earthquakes spoken of in *Phæn.* XI.

Concl. XVIII. And since it appears from *Kircher*'s experiment cited at the latter end of *Obs.* I.

 that

that the communication which ſubterraneous caverns have one with another, is frequently by long extended canals, what wonder is it that earthquakes are ſometimes propagated to very great diſtances, in various directions? as we have obſerved in *Phæn.* XII.

Concl. XIX. But countries whoſe ſoil is ſandy or loamy, are alſo frequently viſited by tremors and ſhocks: Now it is extremely difficult to conceive, how, in ſuch a contexture of earth, any caverns and canals of communication can poſſibly ſubſiſt. This however muſt be underſtood to take place by a kind of conſent of parts, the impulſe being begun at a great diſtance, and the jar propagated by contiguity of ſolid parts, as for example,

——plauſtris concuſſa tremeſcunt
Tecta, viam propter, non magno pondere, tota:
Ferratos utrinque rotarum ſuccutit orbes, &c.

as *Lucretius* elegantly deſcribes it; and *Kircher* aſſents [o]; which ſatisfies for the beginning of *Phæn.* XII.

Concl. XX. The cauſes why mountains and maritime places are moſt obnoxious to ſhocks and ſubverſions, are, firſt, the redundancy of inflammable ſubſtances under mountains, according to *Obſ.* III. and, ſecondly, the winds and blaſts excited by the alliſion of waves, as being great promoters of accenſion, according to *Obſ.* II. But in marſhy and watery places, tho' much abounding in combuſtible matter under ground (ſuch as

[o] Lib. iv. mund. ſubterr. ſect. ii. cap. 10. in fine.

Tuſcany,

Tuſcany, which *Kircher* gives for an example [p]) and this actually ſet on fire, or juſt ready to be ſo, is eaſily quenched by the neighbouring moiſture; ſo that earthquakes cannot be frequent here. And thus have we the cauſe of *Phæn.* XIII. with which compare *Concl.* II. At the ſame time we have the cauſe of the late diſaſter at *Hall*, a ſoil richly impregnated with ſalt; and the ſame inference may be made from what was ſaid in *Obſ.* III. about the iſle of *Ormus*, and ſo we come at the cauſe of *Phæn.* I.

Concl. XXI. The inflammable ſubſtances we have all along been ſpeaking of are not more liable to accenſion in ſpring or ſummer than in autumn and winter, nor more under one conſtellation than another, (*Phæn.* XIV.) It is not therefore ſtrange that no times and ſeaſons have been abſolutely without the related effects. It is however not improbable that the winds blowing ſtronger, or the ſeas running higher at a certain ſeaſon, may have ſome ſhare in promoting them.

Concl. XXII. And as ſmoaky, nitrous and ſulphureous ſteams, before their accenſion, or after it, may eaſily penetrate to the ſources of ſprings; and as aſhes and ſoot are frequently ejected in great quantities, without flame, through clefts and openings of mountains up into the air, the reaſon of *Phæn.* XVI. muſt be very obvious.

Concl. XXIII. Nor is it in the leaſt ſtrange, that miſchievous and venomous exhalations (*Obſ.* VIII.) ſhould, by infecting the air, often bring on peſtilential diſeaſes, as was remarked at the end of the ſame *Phæn.*

[p] Loco ante cit.

Concl. XXIV.

Concl. XXIV. But whether they portend ſeditions and other evils independent of truly natural cauſes,is not the buſineſs of the preſent enquiry. This it is manifeſt from experience, that *Ætna*, *Veſuvius*, &c. do render the circumjacent country extremely fertile by their eruction of a pinguious matter; and that *Greenland* and *Iſland*, otherwiſe intolerably cold, are cheriſhed merely by theſe ſubterraneous fires, and rendered habitable; to ſay nothing of the profit that redounds to the inhabitants from the ſale of the vaſt quantities of ſulphur, wherewith they conſtantly ſupply them, affording them a very comfortable ſupport, which otherwiſe they muſt be altogether in want of.

Concl. XXV. It cannot be queſtioned, but as the waves of the ocean do wear rocks, and waſh away ſhores and the walls of cities; ſo may the waters have free power of waſhing, and excavating the inward parts of the earth in certain places, inſomuch as to cauſe the vaulted roof above to drop in through its own weight; which particular is taken notice of by *Seneca* [q], and has been confirmed by a late example in *Bulgaria*, where a tract of land ſix miles long, ſunk down, without any earthquake, into a deep abyſs; and not long ſince the gazettes mentioned a thing of the like kind of a mountain in *Ruſſia*, where nothing of a ſubterraneous fire would have been ſuſpected, had it not being accompanied with bellowings and roarings: and I wiſh I may be miſtaken in my prognoſtic, as to the town of *Panama* near the iſthmus of *Darien*, on the weſtern coaſt of *America*, ſuffer-

[q] Lib. vi. quæſt. nat. cap. 7.

ing the like fate, especially if what is reported by some be true, that the waves of the sea are frequently heard to roar under the streets.

Concl. XXVI. But whether a tremor, properly so called, may be produced by a violent fall of waters into a subterraneous cavern, let the reader judge from what has been said above, compared with what *Athanasius Kircher* relates [r], that "*At* Panama, *a town of* America, *the flux of the sea is at some times so violent, that the place is full of water, and at the same time an earthquake is felt, and horrid bellowings are heard from under ground.*" And indeed although the hypothesis of *Democritus*, which may be met with at large in *Plutarch*, *Seneca* and *Aristotle*, that subterraneous waters are the cause of earthquakes, be insufficient to solve many of the phænomena, yet it must be acknowledged not to be in all respects absurd.

Concl. XXVII. Nor ought we to oppose *Aristotle*, and others of the ancients, as to the violence of flatuses, especially in a state of rarefaction (*Obs.* VI.) if they could but assign a cause either of instantaneous rarefaction, as that, for example, of air condensed in wind-guns, or of any violent impulse impress'd by continued flatus's from a considerable distance; without which (by the same *Obs.* VI.) the varieties of earthquakes cannot be accounted for (nor indeed the other phænomena, especially the VIIth, if the origin of such impulse be supposed far distant) nor the artificial earthquake of *Arthmesius* described by *Agathius* [s], gain any credit.

[r] Mund. subterr. tom. 1. p. 145. [s] Lib. v.

Concl. XXVIII. Wherefore as to theſe, and other opinions of the ancients, we muſt, in the general, agree with *Seneca* [t], that "*although they are rude and deſtitute of perfection, yet ſtill ought we to excuſe them; and think ourſelves in ſome meaſure indebted to them for whatever improvements we may happen to make.*" As thoſe who broke the ice, and firſt attempted ſuch profound inquiries, in which they would beyond all doubt have ſucceeded, if artillery and gunpowder had been known in their times; for by this alone the moderns were led, and as it were forced into the diſcovery of the cauſes we have here aſſigned, of ſo intricate a matter; of which I will take upon me to produce unqueſtionable proof.

POST-

POSTSCRIPT.

I. I Intend in this additional paper to give due satisfaction to such as would choose to rely on the authorities of other men, rather than trust to their own judgment: And also to prove the truth of a proposition of the utmost importance in the whole science of nature. As to the former, I shall not, as I might, insist that many ancient philosophers deduced the causes of earthquakes, tho' not altogether satisfactorily, from the violent action of fire; and that among the several notions of *Epicurus* on the subject, this was his favorite one, "that earthquakes are produced by some "spiritual flatus converted into fire, which like "thunder, makes havock with whatsoever it meets "in its way," as *Seneca* reports [u]; I rather choose to cite the authorities of a few of the most celebrated moderns.

II. *Caspar Schottus* in his explanation of the nature and action of mines in sieges, says [w], "The "military architects do hollow out a winding nar-"row passage, by the help of a magnetical com-"pass, from the place where the siege is carried "on, to the very fortress they design to demolish; "and under it they work a vault, and close it up "with a door, which has a small hole bored at "its bottom; from which all along, as they re-"tire, they lay a match or train, and set fire to

[u] Lib. toties citat. [w] Mechan. hydraul. p. 61.

"it

" it when they are got out: Thus the whole
" quantity of powder in the vault is kindled at
" once, and the rarefied flame enduring no con-
" finement, blows up the pile that ſtands over it,
" and in an inſtant ſpreads death and terror a-
" round."—After which he adds the following remarkable words. " Nothing was ever deviſed to
" exhibit ſo perfect a reſemblance of an earth-
" quake, as that apparently is no other than an
" effect of rarefaction, and nature in producing
" thoſe concuſſions, operates in a quite ſimilar
" manner; for a flame from ſome ſubterranean
" furnace creeps along a vein of nitre or ſulphur,
" till it arrives at a place where a much larger
" ſtore of thoſe materials are congeſted; which
" being ſuddenly kindled and rarefied, endea-
" vouring to expand itſelf into a larger ſpace,
" ſhakes or overſets the incumbent maſs.

III. How exactly the pupil and his maſter agree, may be ſeen in *Kircher*'s writings [x], where having advanced what we have cited about mines and ordnance at the beginning of *Obſ.* V. he immediately adds,—" who can be ignorant that earthquakes
" have the like origin? They are brought about,
" as has before been ſhewn, in the bowels of
" the earth, and that in the following manner.
" When the powerful effort of ſubterraneous fire
" has broken through the ſides of the caverns of
" mountains, and ſpread itſelf into a large ſpace;
" the air there is put into a violent agitation, and
" the combuſtible particles with which it is copi-
" ouſly impregnated, being ſuddenly kindled, ex-

[x] Mund. ſubterr. lib. iv. p. 221.

" halations are formed in vaſt quantities, and for
" want of a vent for them to eſcape at, the utmoſt
" colluctations enſue which nature is able to en-
" dure; the hollow ſides and vaults of mountains
" are ſhaken, and the ſuperficial parts of the
" earth are lifted up, and, *mark the words*, theſe
" elaſtic vapours work the very ſame effects, as
" gunpowder in artificial mines: They burſt
" through every thing, overſet cities and caſtles,
" form horrid gulphs and new lakes, leaving be-
" hind them the various monuments of deſolation
" and calamity, deſcribed in hiſtorians."

IV. *Deſcartes* goes further [y]. " The ſubtile
" particles of exhalation, ſays he, being too much
" agitated to be converted into oil, when acci-
" dentally driven in any conſiderable quantity
" through the crannies, and into the cavities of
" the earth, do there conſtitute greaſy thick fumes,
" not unlike thoſe which ariſe from a new extin-
" guiſhed candle; and then if any ſpark of fire
" happens to be excited in thoſe cavities, the
" fumes are preſently kindled up, and in conſe-
" quence of an inſtantaneous rarefaction, do ſhake
" the walls of their priſon with prodigious force,
" eipecially if a great deal of ſpirit or *aura* be in-
" termixed with them; and in this manner are
" earthquakes produced." See likewiſe his other opinions about the duration of vulcano's and earthquakes, which I cannot but think highly probable.

V. But the learned *Gaſſendus* of all others, has the moſt ingeniouſly deduced the cauſes of earth-

[y] Princip. part iv. Num. LXXV.

quakes

quakes from ſubterraneous fires, and ſhewn the ſtrict ſimilitude between the effects of artificial mines and earthquakes [z]; the paſſage is ſomewhat prolix, but well worth tranſcribing.—" It ſeems " then much more likely that an earthquake " ſhould be the conſequence of a ſudden inflam- " mation of ſulphureous and bituminous ſteams, " taking fire from an intermixture of nitre, in " ſubterraneous caverns not far below the ſurface " of the ground, it having been before obſerved " that a like ſteam within a cloud, kindles into " lightning. The violent nature of flame, in its " firſt formation, when generated from ſuch ma- " terials, may be ſufficiently known, by attend- " ing to the effects of that of gunpowder fired in " pieces of ordnance; or rather, in military mines, " where the expanſive power of the flame is able " to lift up the weight of a fortreſs or caſtle, and " give a terrible concuſſion to the ſtrongeſt build- " ings in its neighbourhood. Since then a ſmall " quantity of flame let looſe from a ſmall mine, " in compariſon of the maſs of building over it, " is capable of producing ſo great effects, what " may not a far more copious flame in a large " ſubterraneous cavern do to the earth and moun- " tains over it and about it? As the flames of " mines operate with a various ſucceſs, as the " mines are more or leſs confined, greater or leſ- " ſer, deeper or ſhallower, and according to the " cloſeneſs and looſeneſs, dampneſs and dryneſs of " the powder, *&c.* ſometimes producing no ef- " fect at all, ſometimes a ſhock only, and at o-

[z] Animad. in lib. x. Diog. Laert. p. m. 1045 & ſeqq.

 thers

" thers the expected execution; so the flames
" kindled from subterraneous exhalations, ac-
" cording to the various circumstances of the ca-
" verns and vaults, may perhaps often have no
" sensible influence above the surface, either from
" the laxness of the earth, through whose spira-
" cles they may gradually escape and be dissipat-
" ed; or their utmost consequence may be only a
" slight shock or tremor, the incumbent weight
" being too great to be removed; in which case
" the flames will be reflected back, and find a
" passage through some lateral spiracles of the ca-
" vern: Or when the resistance above is great,
" and they cannot otherwise escape, they may oc-
" casion subversions, absorptions, *&c.* Or lastly,
" having forced a sufficient aperture, they may
" belch out fire and ashes, or eject sparrs, mine-
" rals, pumice stones, and fragments of rocks,
" *&c.* partly calcined, and partly melted."

VI. And lastly, let us hear an evidence out of the Peripatetic school, the famous *Andreas Cæsalpinus*, who after having spoken of subterranean exhalations, adds[a], " If at any time a good quan-
" tity of such a substance should be sublimed in-
" to any of the regions of the earth, whose cavi-
" ties are filled with air, and not with water; it
" may easily be set on fire, as happens in the
" clouds. Hence come fiery eruptions in many
" places; hence shocks of earthquakes, and
" oftentimes subversions, when the pores of the
" earth are not open enough to favour the escape

[a] Lib. iii. peripatet. quæst. ix sub finem.

of

"of the generated blaſt: hence ſulphureous beds "and hot ſprings: For ſulphur, bitumen, and "ſuch like inflammable bodies have their origin "from concreted exhalations, which having ac-"quired the igneous principle, do adminiſter to "the duration of ſubterraneous fires; and when "the circumambient bodies become warmed by "ſuch fires, the waters which glide over them are "heated alſo." Which expreſſions, tho' ſome-what obſcure in compariſon of the brighter truths delivered above, yet conſidering them the offspring of the *Latin* Peripatetic ſchool, muſt be allowed to ſhine in ſome degree.

VII. I come now to the other point I propoſed: to enforce it as a ſerious truth, that the ſtupendous effects of earthquakes, whether we conſider them with regard to their immenſe greatneſs or their variety, can be no other than a work every way adequate to the infinite power of the ſupreme Being: And this perhaps may be the more eaſily aſſented to, if an intervention of certain active forces ſubordinate to the ſame divine power, can be demonſtrated; and indeed could nothing of this kind be demonſtrated, it would be a kind of ſacrilege to attempt to aſcribe effects worthy of the divine power and virtue alone, to any natural agent, although ſubordinate to that divine power. For what muſt ſuch agent be? What ſuch virtue ſubordinate and contradiſtinct to the divine power? You will anſwer perhaps without much heſitation, that it muſt be ſubterraneous fire; the efficacy of which is but too apparent and obvious to all who have the misfortune to be placed within the ſphere of its fury.

VIII. I do not deny the ſtupendous energy and power which reſides in fire, the moſt amazing of all God's creatures! It is manifeſt to the moſt vulgar eyes and the dulleſt ſenſes, even to thoſe of brutes. But when I ſurvey it with the philoſophical eye of ſound reaſon, the immediate gift of the divinity, I am plainly convinc'd that its wonderful efficacy is no other than the very efficacy of the divine virtue alone. I have in another place analytically inveſtigated the nature of fire, and found it to conſiſt of two very ſubtile parts, but the one far leſs ſubtile than the other. The leſs ſubtile is made up of ſelect rigid and acute particles of the terrene element, which are abſolutely inert and paſſive; the others of the firſt element, are inconceivably more ſubtile, and extremely moveable; and theſe, in vertue of their perpetual activity, ſet the others in motion, and in this manner produce the univerſally viſible and palpable power of fire. But ſhould we go further, and enquire from whence this rapid agitation of the ſubtile particles of the firſt element is derived? It would be abſurd to ſay they derived it from themſelves, and equally abſurd to ſuppoſe, that theſe, being prime particles, had it from others, prior to themſelves; which if granted, the difficulty would be ſtill the ſame, *&c.* The certain concluſion then muſt be, that the particles of the firſt element did not only once receive that actual mobility which is manifeſt in fire, from an incorporeal principle, prior and ſuperior to all matter, but that it is likewiſe, through

through the perpetual aid of the ſame principle, that it is kept in conſtant poſſeſſion of the ſame. Or, to expreſs the thing more plainly, that the moſt ſubtile components of fire, primarily agitated by the divine will, do, by the ſame divine will, agitate the leſs ſubtile ones, and impel them againſt groſſer bodies; and ſo have all the ordinary and viſible effects of fire hitherto been, now are, and hereafter will be produced: In a word, that the power which we conſider as proper to fire, is in reality the conſtant will of the Deity, whereby he was once pleaſed, that the moſt ſubtile, and by their means alſo the leſs ſubtile parts of fire ſhould be kept in perpetual motion, and that by the mediation of both, all the effects of fire ſhould enſue; and therefore that it will be in vain to imagine that there is any virtue, ſubordinate to the Deity, that can any ways move, or operate upon, the parts of fire, but this divine one alone.

IX. The moſt ſubtile parts of fire are then agitated merely by the divine will; and by them the groſſer *ſpiculæ* of the ſame body: And, by means of the *ſpiculæ*, rapidly impelled on yet groſſer bodies, they are kindled, melted, calcin'd, and burnt to aſhes; and grand maſſes receive impulſes, and are moved in various directions; and all the ſtupendous effects of earthquakes, before related, are brought to paſs. God, according to his good pleaſure, and the eternal order by him eſtabliſhed, makes uſe of various and infinite means (yet of none derived but from himſelf) as paſſive inſtruments, but never employs any other really active vertue, ſubordinate to himſelf. For to what end can ſuch agents exert themſelves? To what can they contribute, when it is his omnipotent virtue alone, that can impreſs upon their ſeveral members the impetus requiſite to the office of their deſtination, whether immediately, or by the intermediation of others, variouſly paſſive, but no ways active, or endued with a virtue contradiſtinct to that of the prime mover? So that it muſt be inſiſted upon again and again, that it is the will of the omnipotent Creator alone, that acts and moves, and by moving governs and regulates all things in the univerſal world, and that immediately, in conſequence of a proper virtue: That is without the intervention of any other active virtue of any creature whatever, though indeed mediately, in regard to the action of that only divine virtue, with reſpect to the diſpoſition, aptitude, and capacity of various recipients.

X. But could not the great Creator of the univerſe communicate to fire an active power of burning, *&c.* in vertue of which it might afterwards perform all its uſual effects, and, of itſelf, bring on earthquakes?

I anſwer: I know very well that with many, this is the main obſtacle which hinders their aſſent to the philoſophical truth which I contend for, tho' clear enough in itſelf. I am, I own, very deſirous of ſhewing the impoſſibility of communicating ſuch active forces to ſubſtances merely corporeal. I intreat therefore my readers attention to what I have already ſaid, as well as to what I am going to ſay concerning the power of fire to burn, *&c.*

Unleſs

Unleſs I were to expreſs the definition of fire in abſtract terms, inſtead of conſidering it under any real agitation and motion, obvious partly to the ſenſes, and partly to the imagination, I muſt be obliged, with moſt modern philoſophers, to ſuppoſe a twofold motion, one of the groſſer and terrene particles piercing, cutting, breaking and diſſolving the continuity of other bodies, and inflicting the moſt exquiſite pains on ſenſitive bodies; and another, of the inconceivably ſubtile parts, ſwiftly pervading in all directions the pores of all bodies, not previouſly occupied by themſelves in conſort with their terrene *ſpiculæ*, and that not only without any detriment, but even ſenſible perception. It is clear and manifeſt that the impetus of the former particles, ſince it is paſſively dependent on the ſuppoſed ſwift agitation of the latter, cannot conſtitute any active power in fire. Wherefore, if there were any active power at all in fire, it muſt be aſcribed to the agitation of its very ſubtile parts (ſuppoſing it has none more ſubtile ſtill than thoſe, *&c.* to do which would be weak and abſurd,) which is the ſame as to ſay it is communicated to it by God himſelf. Now ſuch agitating force could be communicated to it no otherwiſe than either by giving to the particles a power of agitating themſelves (which is abſurd to all ſound reaſon, and even to the Peripatetics themſelves) or by willing that they ſhould be ſo agitated. But ſince to be agitated implies ſomething paſſive, and in this inſtance, dependent inevitably on the divine will; it is manifeſt that in fire there is no active power, properly ſo called, beſides the ſole efficacy

efficacy of the divine will, whereby that more ſubtile part of it communicates motion to the groſſer particles, impelling them upon other bodies, and ſo producing other conſequent effects; in reſpect of which, the motion of the *ſpiculæ* may indeed be called active, as alſo that of the ſubtile parts in reſpect of the motion of the *ſpiculæ*, though they are all of them abſolutely paſſive. It is then the perfection of the divine power alone, not to ſtand in need of any intrinſic motive power, and as ſuch it is abſolutely and truly active, and efficiently productive of the motions eſſential to fire, and of innumerable others thereon depending.

OF

OF THE NATURE OF EARTHQUAKES.

More particularly of the Origin of the Matter of them, from the Pyrites alone.

I Have elſewhere [a] ſhewn that the breath of the pyrites is ſulphur *ex tota ſubſtantia:* alſo that it naturally takes fire of itſelf. Again that the material cauſe of thunder and lightning, and of earthquakes, is one and the ſame; *viz.* the inflammable breath of the pyrites. The difference is, that one is fixed in the air; the other under ground: of which laſt, theſe I think are ſufficient arguments. A thing burnt with lightning ſmells of very brimſtone; again, the ſubtilty and thinneſs of the flame; alſo the manner of its burning, which is often obſerved to be *particulatim*, or in ſmall ſpots, vapour-like. And of earthquakes,

[a] *De fontibus medicatis Angliæ.*

the

the ſulphureous ſtink of waters ſmelt before, and of the very air itſelf after them; of which innumerable inſtances occur in the relations of them.

They alſo agree in the manner of the noiſe, which is to be carried on, as in a train fired; the one rolling and rattling through the air, taking fire as the vapours chance to drive; as the other fired under ground, in like manner moves with deſultory noiſe, as it ſhall chance to be continued.

That the earth is more or leſs hollow, is made probable by what is found every where in mountains, *viz.* natural cavities or chambers, which the miners of the north call *ſelf-opens.* Theſe they meet with very frequently, ſome vaſtly great, and others leſs, running with ſmall ſinus's. And I doubt not, upon diligent inquiry, a great catalogue of ſuch might be had, diſcovered in the memory of man: beſides many there are, which are known to be open to the day, and to diſcover themſelves without digging, as *Pool's Hole*, *Oakie Hole*, &c. Again, the great and ſmall ſtreams, which do ariſe from under the mountains, do evince the hollowneſs and ſinuouſneſs of them. Add to theſe, that many ſinus's are made in that inſtant, and are continued by the exploſion and rending of the firſt matter fired; which may, and do very probably, cloſe again, when the force of that exploſion is over; but are ſufficiently open to continue the earthquake.

That theſe ſubterraneous cavities are at certain times, and in certain ſeaſons full of inflammable vapours, the damps in our mines ſufficiently witneſs; which fired, do every thing as in an earthquake, ſave in a leſſer degree.

Now, that the pyrites alone (which is our preſent taſk) of all the known minerals, yields this inflammable vapour, I think is highly probable for theſe reaſons.

I. Becauſe no mineral or oar whatſoever is ſulphureous, but as it is wholly, or in part a pyrites: and although this does contradict the general opinion of the chymiſts; yet they muſt excuſe me if I diſſent from them in this particular: for where any of them ſhall find me brimſtone naturally contained in an ore; there, I am very forward to believe, I ſhall find them iron alſo, by the loadſtone; ſo that betwixt us we ſhall have diſcovered the pyrites diſguiſed in that ore or mineral. I have carefully made the experiment in very many of the foſſils of *England*, and do find them all to contain *iron*, wherever *brimſtone* is, as I have elſewhere declared.

II. Becauſe there is but one ſpecies of brimſtone, that I know of, at leaſt with us in *England:* And ſince the pyrites naturally and only yields it, it is but reaſonable, wherever brimſtone is found, though in the air, or under ground in vapour, to think that that alſo proceeds from it.

If it be objected, that there is a ſulphur vive, or natural brimſtone, which is no pyrites; I anſwer, that I am not willing to grant this, but do take all pure ſulphur to have been once produced by the fire: for what is found in and about the burning mountains, is certainly the effects of ſublimation: and thoſe great quantities of it, ſaid to be found

found about the ſkirts of volcano's, is only an argument of the long duration and vehemence of thoſe fires.

If it be further objected, that the ſulphur vive indeed, or ruff brimſtone, as they call it, had from *Hecla* and *Italy*, is opaque, and agrees not with the tranſparent and amber-like ſulphur vive of the ancients, ſo that the miſtake is in the druggiſts, that we have not right natural brimſtone; I reply, that grant the difference, yet it does not follow, that that alſo was produced by ſublimation, no more than that the ſtalactites, or water-wrought ſtone, is not ſo made, for that ſome of it is found opaque, and ſome chryſtalline.

But this we will grant; that poſſibly the pyrites of the volcano's or burning mountains may be more ſulphureous than ours. And indeed it is plain, that ſome of ours in *England* are very lean, and hold but little ſulphur; others again very much.

And this may be one reaſon, why *England* is ſo little troubled with earthquakes; and *Italy* and almoſt round the *Mediterranean Sea* ſo very much.

Another reaſon is, the paucity of pyrites in *England*; where they are indeed, ſome little in all places, but moſtly, *ſparſim*; and if perchance in *beds*, thoſe are comparatively thin, to what probably they were in the burning mountains, as the vaſt quantity of ſulphur from thence ſublimed, doth ſeem reaſonably to imply. Alſo if we compare our earthquakes, and our thunder and lightning with theirs; there it lightens almoſt daily, eſpecially in ſummer time, here ſeldom; there thunder and lightning is of long duration, here ſoon over; there the earthquakes are frequent, long and terrible, with many paroxyſms in a day, and that for many days; here very ſhort, a few minutes, and ſcarce perceptible. To this purpoſe the ſubterraneous cavities in *England* are ſmall, and few, compared to the vaſt vaults in thoſe parts of the world; which is evident from the ſudden appearance of whole mountains and *iſlands*.

If yet it ſhall be inſiſted upon, that there are other inflammable minerals beſides the pyrites; we grant there are ſo, but, by the providence of God, not to be found in *England*, that I know of, and not in any quantity in any place of the world, that I can learn; which is well for mankind, becauſe they are very poiſons, as the orpiments; but they are all ſpecifically diſtinct from brimſtone, which, as we have ſhewn, no ore yields but iron; ſo that *Nero* (as *Pliny* teſtifies, who was of his time and his court) cauſed them to be wrought in quantity, but they would not turn to account. And, by the by, ſome authors have aſſigned this as a good reaſon, againſt any medicine that ſhall be made out of gold, as fond as we are of an *aurum potabile*, as having naturally a deleterious quality: but this is beſides my purpoſe.

Of the ſpontaneous firing of the Pyrites.

IF it ſhall be objected, that no body is kindled by itſelf: I anſwer, that it ſeems to me apparently otherwiſe; for that vegetables will heat, and take fire of themſelves, as in the frequent inſtance

ſtance of wet hay; and animals are naturally on fire, and a man doth then ſufficiently demonſtrate it when he is in a fever. But amongſt minerals, the pyrites, both in groſs and in vapour, is actually of its own accord fired. Dr. *Power* has actually recorded at large in his *Micrographia* [b], a famous inſtance of it; and the like not very rarely happens. And that damps naturally fire of themſelves, we have the general teſtimony of miners and of the ſame author [c].

Again, the volcano's all the world over, argue as much: for we, with great probability, believe them to be mountains made up in great part of pyrites, by the quantities of ſulphur thence ſublimed, and the application of the loadſtone to the ejected cinder. I go further.

That theſe volcano's were naturally kindled of themſelves, at or near the creation, is probable, becauſe there is but a certain known number of them, which have all continued burning beyond the memoirs of hiſtory: few or none of them, that I know of, have even totally decayed or been extinct, unleſs poſſibly by the ſubmerſion of the whole; being abſorb'd into the ſea: though they do indeed burn more fiercely ſometimes, than at others, for other reaſons. So that it ſeems to me as natural to have actual fire in the terreſtrial world from the creation, as to have ſea and water.

Again, if theſe volcano's did not kindle of themſelves, what cauſe can we imagine to have done it? Of the ſun; we anſwer, *Hecla* placed in ſo extreme cold a climate, was kindled, for ought

[b] *Power Microg.* p. 61. [c] *Id.* p. 181.

I

I can ſee by the natural hiſtory of both, as ſoon as *Ætna*, or *Fuegos*, or the moſt ſoutherly. Not the accidents happening from man; for if man was, as we muſt believe, created ſolitary and topical, they were none of his kindling, becauſe they ſeemed to be fired before the world was overpeopled: beſides, they are moſtly the very tops of vaſt high mountains, and therefore the moſt unfit for the habitation of man.

If we ſay lightning and thunder, and earthquakes, we beg the queſtion; for the cauſe of one is the cauſe of the other, and they are one and the ſame.

It remains therefore, very probably, that they were kindled of themſelves.

I for my part know no ſubject in the whole mineral kingdom ſo general and laſting for the fuel of theſe mountains, as the pyrites; which I have ſaid alone to yield ſulphur, and naturally reſolves itſelf into it, by a kind of vegetation.

About the durable burning of the pyrites, theſe are inſtances. *Scotch* coal hath leſs of the pyrites in it, being moſtly made up of coal bitumen, and therefore it burns and conſumes quickly, and leaves a white cinder. Sea-coal, or that coal which comes from *Newcaſtle* by ſea to us, and for that reaſon ſo called, burns ſlowly; and the *Sunderland* ſea-coal ſo ſlowly, that it is ſaid by proverb, *to make three fires*: this hath much pyrites mixed with it, and burns to a heavy rediſh cinder, which is iron, by the magnet. But I have ſeen, and have a ſpecimen by me of a coal from *Ireland*, the proprietor of the pits is Sir *Chriſtopher Wandsford*,

 which

which is ſaid to be ſo laſting, that it will continue twenty four hours red hot, and almoſt keep its figure. This ſeems to be in a great part pyrites by the weight and colour.

There are two ſorts of inſtances, beſides the arguments I have already urged, which to me are alone ſufficiently convincing, and very much favour the opinion I have offered; that thunder and lightning owe their matter to the ſole breath of the pyrites. And although I am as loth, and as backward as any man, to give credit to ſuch inſtances, which ſeem rather prodigies, than the phænomena of nature; yet becauſe they often occur in hiſtory, it is at leaſt fitting to bring them under further inquiry and examination, that if they can be confuted as falſe, ſo much may be done for poſterity; and that we at leaſt may not leave upon our regiſters matters of fact not true, if they can be fairly ſet aſide.

The firſt ſort of them are thoſe which tell us of iron to have fallen in great maſſes, and alſo in powder, after the manner of rain, out of the air.

In a part of *Italy* it rained iron in ſuch a year, and in *Germany* a great body of iron-ſtone fell at ſuch a time: The like *Avicenn* affirms. *Julius Scaliger* ſays he had by him a piece of iron which was rained in *Savoy*, where it fell in divers places. *Cardan* reports 1200 ſtones to have fallen from heaven, and one of them weighed 120 pounds, ſome of them 30 pounds, ſome 40, very hard, and of the colour of iron.

Now, that which is very remarkable, ſays *Gilbert*, where thoſe inſtances are reckoned up, and

a

a very probable argument for the truth of ſuch like inſtances, is, that it is no where recorded, that it ever rained gold or ſilver ore, or tin or lead; but copper hath been alſo ſaid to have fallen from the clouds.

And here I muſt note by the by, that wherever the pyrites is mentioned by the ancients, it is always to be underſtood of the *copper pyrites*; they ſcarce having had any knowledge of the iron pyrites: And therefore the raining of copper makes it yet more probable, becauſe of its great affinity with iron, which I ſhall have occaſion on ſome other time to diſcourſe of.

Now this *Ferrum* or *Æs Nubegnum*, if there was ever any ſuch, was concreted of the breath of the pyrites, which we have elſewhere ſhewn to be the ſulphur *ex tota ſubſtantia*.

The other inſtance, which I ſay is owing to our regiſters, is of lightning being magnetic [d].

This I am ſure of, I have a petrified piece of aſh which is magnetic; that is, the pyrites *in ſucco*; which makes it probable it may be magnetic alſo in vapour.

[d] Philoſoph. Tranſact. N° 127.

DISCOURSES

CONCERNING

EARTHQUAKES.

Vidi ego quod fuerat quondam ſolidiſſima tellus
Eſſè fretum; vidi fractas ex littore terras;
Et procul a pelago chonchæ jacuere marinæ,
Et vetus inventa eſt in montibus anchora ſummis;
Quodque fuit campus, vallem decurſus aquarum
Fecit, et eluvie mons eſt deductus in æquor.
Ovid. Metam. Lib. xv.

PROPOSITIONS.

I. THere are found in moſt countries of the earth, and even in ſuch where it is ſomewhat difficult to imagine, by reaſon of their vaſt diſtance from the ſeas or waters, how they ſhould come there, great quantities of bodies, reſembling both in ſubſtance and ſhape, the ſhells of divers ſort of ſhell-fiſhes; and many of them ſo exactly, that any one that knew not from whence they came, would without the leaſt ſcruple, firmly believe them to be the ſhells of of ſuch fiſhes: but being found in places ſo unlikely to have produced them, and not conceiving how elſe they ſhould come there; they are generally believed to be real ſtones formed into thoſe ſhapes, either by ſome plaſtic virtue inherent in thoſe parts of the earth, which is extravagant enough, or elſe by ſome celeſtial influence or aſpect of the planets operating at a diſtance upon the yielding matter of the parts of the earth, which is much more extravagant. Of this kind are all thoſe ſeveral ſorts of oyſter-ſhells, cockle-ſhells, muſcle-ſhells, periwinkle-ſhells, *&c.* which are found in *England*, *France*, *Spain*, *Italy*, *Germany*, *Norway*, *Ruſſia*, *Aſia* and *Africa*, and divers other places; of which we have very good teſtimony from authors of good credit.

II. There often have been, and ſtill are daily found in other parts of the earth, buried below the preſent ſurface thereof, divers ſorts of bodies, beſides ſuch as I newly mentioned, reſembling both in ſhape, ſubſtance, and other properties, the parts of vegetables, having the perfect rind or bark, pith, pores, roots, branches, gums, and other conſtituent parts of wood; and though in another poſture, lying for the moſt part horizontal, and ſometimes inverted, and much different from that of the like vegetables when growing; and wanting alſo for the moſt part, the leaves, ſmaller roots and branches, the flower and fruit, and the like ſmaller parts, which are common to trees of that kind: of which ſort is the *lignum foſſile*, which is found in divers parts of *England*, *Scotland*, *Ireland*, and various parts of

Italy, *Germany*, the *Low-Countries*, and indeed almoſt in every country of the world.

III. There are often found in divers other parts of the earth bodies, reſembling the whole bodies of fiſhes, and other animals and vegetables, or the parts of them, which are of a much leſs permanent nature than the ſhells abovementioned; ſuch as fruits, leaves, barks, woods, roots, muſhrooms, bones, hoofs, claws, horns, teeth, *&c.* But in all other properties of their ſubſtance, ſave their ſhape, are perfect ſtones, clays or earths, and ſeem to have nothing at all of figure in the inward parts of them. Of this kind are thoſe commonly called thunder-bolts, helmet-ſtones, ſcrew-ſtones, wheel-ſtones, *&c.*

IV. The parts of the earth in which theſe kinds have been found, are ſome of them ſome hundred of miles diſtant from any ſea, as in ſeveral hills of *Hungary*, the mountain *Taurus*, the *Alpes*, &c.

V. Divers of thoſe parts are many ſcores, nay ſome many hundreds of fathoms above the level of the ſurface of the next adjoining ſea, they having been found in ſome of the moſt inland, and on ſome of the higheſt mountains in the world.

VI. Divers other parts where theſe ſubſtances have been found, are many fathoms below the level both of the ſurface of the next adjoining ſea, and of the ſurface of the earth itſelf, they having been found buried in the bottoms of ſome of the deepeſt mines and wells, and incloſed in ſome of the hardeſt rocks and tougheſt metals. Of this we have continual inſtances in the deepeſt lead and tin-mines, and a particular inſtance in the well

dug

dug in *Amſterdam* ; where at the depth of 99 feet was found a layer of ſea ſhells mixed with ſand, of four feet thickneſs ; after the diggers had paſſed through ſeven foot of garden-mould, nine foot more of black peat, nine foot more of ſoft clay, eight of ſand, four of earth, ten of potters clay, four more of earth, ten foot more of ſand, upon which the ſtakes or piles of the *Amſterdam* houſes reſt ; then two foot more of potters-clay, and four of white-gravel, five of dry earth, one of mix'd, fourteen of ſand, three of ſandy clay, and five more of potters-clay mixed with ſand. Now below this layer of ſhells, immediately joining to it, was a bed of potters-clay of no leſs than 102 foot thick.

VII. There are often found within the bodies of very hard and cloſe ſtone, as marbles, flints, *Portland* and *Purbeck* ſtones, *&c.* which lye upon, or very near to the ſurface of the earth, great quantities of theſe kind of figured bodies or ſhells; and there are many of ſuch ſtones which ſeem to be made of nothing elſe.

Theſe phænomena, as they have hitherto much puzzled all natural hiſtorians and philoſophers to give an account of them, ſo in truth are they in themſelves ſo really wonderful, that 'tis not eaſy, without making multitudes of obſervations, and comparing them very diligently with the hiſtories and experiments that have been already made, to fix upon a plauſible ſolution of them. For as on the one ſide, it ſeems very difficult to imagine that nature formed all theſe curious bodies for no other end, than only to play the mimick in the

 mineral

mineral kingdom, and only to imitate what ſhe had done for ſome more noble end, and in a greater perfection in the vegetable and animal kingdoms; and the ſtricteſt ſurvey that I have made, both of the bodies themſelves, and of the circumſtances obvious enough about them, do not in the leaſt hint any thing elſe; they being promiſcuouſly found of any kind of ſubſtance, and having not the leaſt appearance of any internal or ſubſtantial form, but only of an external or figured ſuperficies. As, I ſay, 'tis ſomething harſh to imagine that theſe thus qualified bodies ſhould, by an immediate plaſtic virtue, be thus ſhaped by nature, contrary to her general method of acting in all other bodies; ſo on the other ſide, it may ſeem at firſt hearing ſomewhat difficult to conceive how all thoſe bodies, if they either be the real ſhells or bodies of fiſh, or other animals or vegetables, which they repreſent, or an impreſſion left on thoſe ſubſtances from ſuch bodies, ſhould be in ſuch great quantities tranſported into places ſo unlikely to have received them from any help of man, or from any other obvious means.

The former of theſe ways of ſolving theſe phænomena I confeſs I cannot, for the reaſons I now mentioned, by any means aſſent unto; but the latter, tho' it has ſome difficulties alſo, ſeems to me not only poſſible, but probable.

The greateſt objections that can be made againſt it, are *1ſt*, By what means thoſe ſhells, woods, and other ſuch like ſubſtances, if they really are the bodies they repreſent, ſhould be tranſported to, and buried in the places where they are found? And

And 2*dly*. Why many of them ſhould be of ſubſtances wholly differing from thoſe of the bodies they repreſent; there being ſome of them which repreſent ſhells of almoſt all kinds of ſubſtances, clay, chalk, marble, ſoft ſtone, harder ſtone, marble, flint, marchaſite, ore, *&c.*

In anſwer to both which, and ſome other of leſs importance, which I ſhall afterwards mention, give me leave to propound theſe following propoſitions, which I ſhall endeavour to make probable. Of theſe in their order.

I. All, or the greateſt part of thoſe curiouſly figured bodies, found up and down in various parts of the world, are either thoſe animal or vegetable ſubſtances they repreſent, converted into ſtone, by having their pores filled up with ſome petrifying liquid ſubſtance, whereby their parts are, as it were, lock'd up and cemented together in their natural poſition and contexture; or elſe they are the laſting impreſſions, made on them at firſt, whilſt a yielding ſubſtance, by the immediate application of ſuch animal or vegetable body, as was ſo ſhaped; and that there was nothing elſe concurring to their production, ſave only the yielding of the matter to receive the impreſſion, ſuch as melted wax affords to the ſeal: or elſe a ſubſiding or hardning of the matter, after by ſome kind of fluidity it had perfectly filled or incloſed the figuring vegetable or animal ſubſtance, after the manner as a ſtatue is made of plaiſter of *Paris*, or alabaſter duſt beaten, and boiled, mixed with water, and poured into a mould.

II. There

II. There ſeems to have been ſome extraordinary cauſe which did concur to the promoting of this coagulation or petrifaction; and that every kind of matter is not of itſelf apt to coagulate into a ſtrong ſubſtance, ſo hard as we find moſt of thoſe bodies to conſiſt of.

III. The concurrent cauſes aſſiſting towards the turning of theſe ſubſtances into ſtone, ſeem to have been one of theſe; either ſome kind of fiery exhalation, ariſing from ſubterraneous eruptions or earthquakes; or, ſecondly, a ſaline ſubſtance, whether working by diſſolution and congelation, or cryſtallization, or elſe by præcipitation and coagulation; or thirdly, ſome glutinous or bituminous matter, which upon growing dry or ſettling, grows hard, and unites ſandy bodies together into a pretty hard ſtone; or fourthly, a very long continuation of theſe bodies under a great degree of cold and compreſſion.

IV. Waters themſelves may in tract of time be perfectly tranſmuted into ſtone, and remain a body of that conſtitution, without being reducible by any art yet commonly known.

V. Divers other fluid ſubſtances have, after a long continuance at reſt, ſettled and congealed into much more hard and permanent ſubſtances.

VI. A great part of the ſurface of the earth hath been ſince the creation transformed and made of another nature; namely many parts which have been ſea are now land, and divers other parts are now ſea which were once a firm land; mountains have been turned into plains, and plains into mountains, and the like.

VII. Di-

VII. Divers of theſe kinds of transformations have been effected, in theſe iſlands of *Great Britain*; and 'tis not improbable but that many very inland parts of this iſland, if not all, may have been heretofore all covered with the ſea, and have had fiſhes ſwimming over it.

VIII. Moſt of thoſe inland places, where theſe kind of ſtones are, or have been found, have been heretofore under water; and either by the departing of the waters to another part or ſide of the earth, by the alteration of the center of gravity of the whole bulk, which is not impoſſible; or rather by the eruption of ſome kind of ſubterraneous fires or earthquakes, whereby great quantities of earth have been raiſed above the former level of thoſe parts, the waters have been forced away from the parts they formerly covered, and many of thoſe ſurfaces are now raiſed above the level of the waters ſurface, many ſcores of fathoms.

IX. It ſeems not improbable that the tops of the higheſt and moſt conſiderable mountains in the world have been under water, and that they themſelves ſeem moſt probably to have been the effects of ſome very great earthquake, ſuch as the *Alpes* and *Apennine* mountains, *Caucaſus*, the pike of *Teneriffe*, the pike in the *Tercera's* and the like.

X. It ſeems not improbable, but that the greateſt part of the inequality of the earth's ſurface may have proceeded from the ſubverſion and tumbling thereof, by ſome preceding earthquakes.

XI. There have been many other ſpecies of creatures in former ages, of which we can find none at preſent; and 'tis not unlikely alſo but that there may

may be divers new kinds now, which have not been from the beginning.

There are ſome other conjectures of mine yet unmentioned, which are more ſtrange than theſe, which I ſhall defer the reciting of at preſent, becauſe, though I have divers obſervations concurring; yet having not been able to meet with ſuch as may anſwer ſome conſiderable objections that they are liable to, I will rather endeavour to make probable thoſe already mentioned, by ſetting down ſome of thoſe obſervations (for it would be tedious to inſert them all) I have collected both out of authors, and from my own experience.

The firſt was, that theſe figured bodies diſperſed over the world, are either the beings themſelves petrified, or the impreſſions made by thoſe beings. To confirm which, I have diligently examined many hundreds of theſe figured bodies, and have not found the leaſt probability of a plaſtic faculty. For firſt, I have found the ſame kind of impreſſion upon ſubſtances of an exceeding different nature; whereas nature in other of her works, does adapt the ſame kind of ſubſtances to the ſame ſhape: the fleſh of a horſe is differing from that of a hog, or ſheep, or from the wood of a tree, or the like; ſo the wood of box, for inſtance, is differing from the wood of all other vegetables; and if the outward figure of the plant or animal differ, to be ſure their fleſh alſo differs: and under the ſame ſhape you always meet with ſubſtances of the ſame kind; whereas here I have obſerved ſtones bearing the ſame figure, or rather impreſſion, to be of hugely differing natures; ſome of of clay, ſome of chalk, ſome of ſpar, ſome of marble, ſome of a kind of freeſtone, ſome like cryſtals or diamonds, ſome like flints, others a kind of marchaſite, others a kind of ore. Nay in the ſame figured ſubſtance I have found divers ſorts of very differing bodies or kinds of ſtone, ſo that one has been made up partly of ſtone, partly of clay, and partly of marchaſite, and partly of ſpar, according as the matter chanced to be jumbled together, and to fill up the mould of the ſhell.

Another circumſtance which makes this conjecture the more probable, is, that the outward ſurface only of the body is formed, and that the inward part has nothing of ſhape that can reaſonably be referred to it; whereas we ſee, that in all other bodies that nature gives a ſhape to, ſhe figures alſo the internal parts, or the very ſubſtance of it, with an appropriate ſhape. Thus in all kinds of minerals, as ſpars, cryſtals, and divers of the precious ſtones, ores, and the like, the inward parts of them are always correſpondent to the outward ſhape; as in ſpar, if the outward part be ſhaped into a rhomboidical parallelopiped, the inward part of it is ſhaped in the ſame manner, and may be cleft out into a multitude of bodies of the like form and ſubſtance.

Another circumſtance is, that I have in many found the perfect ſhell incloſed, which I have ſometimes been able to take out intire, and found to be, both by its ſubſtance and ſhape, and reflective ſhining, and the like circumſtances, a real ſhell of a cockle, perriwinkle, muſcle, or the like.

And

And further, I have found in the ſame place divers of the ſame kinds of ſhells, not filled with a matter that was capable of taking the impreſſion, but with a kind of ſandy ſubſtance, which lying looſe within it, could be eaſily ſhook out, leaving the incloſing ſhell perfectly intire and empty; others I have ſeen which have been of black flint, wherein the impreſſion has been made only of a broken ſhell, which ſtuck alſo in it, the other part of the ſurface of that ſtone, which was not within the ſhell, remaining only formed, like a common flint.

And, which ſeems to confirm this conjecture, much more than any of the former arguments, I had this laſt ſummer an opportunity to obſerve upon the *ſouth* part of *England*, in a clift whoſe bottom the ſea waſht, that at a good height in the clift above the ſurface of the water, there was a layer, or vein of ſhells, which was extended in length for ſome miles; out of which layer I digged, and examined many hundreds, and found them to be perfect ſhells of cockles, perriwinkles, muſcles, and divers other ſorts of ſmall ſhell-fiſhes; ſome of which were filled with the ſand with which they were mixed; others remained empty, and perfectly intire. From the ſea water's waſhing the under part of this clift, great quantities of it do every year tumble or founder down, and fall into the ſalt water, which are waſhed alſo by the ſeveral mineral waters iſſuing out at the bottom of the clifts. Of theſe foundered parts I examined very many parcels, and found ſome of them made into a kind of hardened mortar, or very ſoft ſtone, which

which I could eaſily with my foot, and even almoſt with my finger, cruſh in pieces: others that had laid a longer time expoſed to the viciſſitudes of the riſing and falling tides, I found grown into pretty hard ſtones; others that had been yet longer, I found converted into a very hard ſtone, retaining exactly the ſhape of the incloſing ſhell: and in the part of the ſtone which had encompaſſed the ſhell, there was left remaining the perfect impreſſion and form of the ſhell; the ſhell itſelf continuing, as yet, of its natural white ſubſtance, tho' much decayed or rotted by time: but the body incloſing and included by the ſhell, I found exactly ſtamp'd like thoſe bodies whoſe figures authors generally affirm to be the product of a plaſtic or vegetative faculty working in ſtones.

Another argument, that theſe petrified ſubſtances are nothing but the effects of thoſe ſhells being filled with ſome petrifying ſubſtances, is this, that among thoſe which are called *Cornua Ammonis*, or ſerpentine ſtones, found about *Keinſham*, and in ſeveral other parts of *England*, and in other countries, as the *Balnea Bollenſia*, which are indeed nothing elſe but the moulding off from a kind of ſhell which is ſhaped much like a *nautilus* ſhell, the whole cavity being ſeparated with divers ſmall valves or partitions, much after the ſame manner as thoſe ſhells of the *nautilus* are commonly obſerved to be. Among theſe ſtones, I ſay, I have, upon breaking, found ſome of the cavities between thoſe partitions remain almoſt quite empty; others I have found lined only with a kind of tartareous, or rather cryſtalline ſubſtance, which has ſtuck to the ſides, and been figured like tartar, but of a clear and tranſparent ſubſtance like cryſtal; whereas others of the cavities of the ſame ſtone, I have found filled with divers kinds of ſubſtances very differing: whence I imagine thoſe tartareous ſubſtances to be nought elſe but the hardening of ſome ſaline fluid body, which might ſoak in through the ſubſtance of the ſhell. Others of theſe I have, which are quite of a tranſparent ſubſtance, and ſeem to be produced from the petrifaction of the water that had filled them: others I have found filled with a perfect flint, both which I ſuppoſe to be the productions of water petrified: and I may perhaps hereafter make it probable, that all kinds of flints and pebbles have no other original.

I could urge many other arguments to make my firſt propoſition probable, that all thoſe curiouſly ſhaped ſtones, which the moſt curious naturaliſts moſt admire, are nothing but the impreſſions made by ſome real ſhell, in a matter that at firſt was yielding enough, but which is grown harder with time. To this very head alſo may be referred all thoſe other kinds of petrified ſubſtances, as bones, teeth, crabbs, fiſhes, wood, moſs, fruit, and the like; ſome of all which kinds I have examined, and by very many circumſtances, too long to be here inſerted, judge them to be nothing elſe but a real petrifaction of thoſe ſubſtances they reſemble.

My ſecond propoſition will not be difficult to prove, that if theſe be the effects of petrifaction or coagulation, it muſt be from ſome extraordinary cauſe; and this becauſe we find not many experiments of producing them when and where we will: beſides

we

we find that moſt things, eſpecially animal and vegetable ſubſtances, after they have left off to vegetate, do ſoon decay, and, by divers ways of putrefaction and rotting, loſe their form and return to duſt; as we find wood, whether expoſed to the air or water, in a little time to waſte and decay, eſpecially ſuch as is expoſed to the alteration of both, and even in thoſe places where theſe petrified ſubſtances are to be met with. The like we find of animal ſubſtances; and we have but ſome few experiments of preſerving thoſe bodies, to make them as permanent as ſtone, and few of making them into a ſubſtance of the like nature.

The third thing therefore, which I ſhall endeavour to ſhew, is, that the concurring cauſes to theſe petrifactions, ſeem to be either ſome kind of petrifying water, or elſe ſome ſaline or ſulphureous mixture, with the concurrence of heat, from ſome ſubterraneous fire or earthquake; or elſe a very long continuation of thoſe bodies under a very great degree of cold, and compreſſion, and reſt. That petrifying waters may be able to convert both animal and vegetable ſubſtances into ſtone, I could, beſides ſeveral trials of my own, bring multitudes of relations out of natural hiſtorians: but theſe are ſo common in almoſt all countries, and ſo commonly taken notice of by the curious, that I need not inſtance. *Camden* and *Speed* will tell you of abundance here in *England*, as the *Peak* in *Derbyſhire*, and in ſeveral other ſubterraneous caverns in *England*. The water itſelf does, by degrees, produce ſeveral conical pendulous bodies of ſtone, ſhap'd and hanging like icicles from the roof

 of

of the vault; and dropping on the bottom, it raiſes up alſo conical ſpires, which, by degrees, endeavour to meet the former pendulous *ſtriæ*. And indeed I have generally obſerved it, that wherever there is a vault made with lime under ground, into which the rain-water, ſoaking through a pretty thickneſs of ground, does at laſt penetrate through the arch: I have in ſeveral places, I ſay, obſerved, that that water does incruſtate the roof with ſtone, and in many places of it generate ſmall pendulous icicles. This water I have found in a little time to incruſtate ſticks, or the like vegetable ſubſtances, with ſtone, and in ſome places to penetrate into the pores of the wood, filling them up with ſmall cylinders of ſtone. This I have obſerved alſo in ſeveral of the arches of St. *Paul*'s church, which have been uncovered and laid open to the rain, though there be no earth for it to ſoak through. And altho' I have never yet been able to petrify a ſtick throughout, yet I have now by me ſeveral pieces, that retain ſo perfectly all the figure of the wood, and are yet ſo perfectly, in all other properties, ſtone, that I find not the leaſt reaſon of doubt to believe, that thoſe pieces have been actual wood; having ſtill the bark, the clefts, the knots, the grain, the pores, and even thoſe too which, for their ſmallneſs, I have elſewhere called microſcopical; tho' I confeſs ſome of thoſe more perfect pieces ſeem to have been petrified from ſome more ſubtile and inſinuating petrifying water, than thoſe I newly mentioned: and 'tis not improbable but that ſome ſubterraneous ſteams and heat may have contributed ſomewhat towards this effect. But firſt I ſhall

endeavour

endeavour to make it probable, that theſe petrified bodies may have been placed in thoſe parts where they are found, by ſome kind of transformation wrought on the ſurface of the earth, by ſome earthquake: and to this end I ſhall by and by mention ſome ſtrange alterations that have been made by earthquakes, after I have firſt made probable my fourth conjecture.

The fourth propoſition therefore to be explained and made probable is, that waters themſelves of divers kinds, are, and may have been tranſmuted perfectly into a ſtony ſubſtance, of a very permanent conſtitution, being ſcarcely reducible again into water by any art yet commonly known: and that divers other liquid or fluid ſubſtances have in tract of time ſettled and congealed into much more hard, fixt, ſolid and permanent forms than they were of at firſt.

The probability of which propoſition may appear from theſe particulars.

I. That almoſt in all ſtreams and running waters there is to be found great quantities of ſand at the bottom, many of which ſands both by their figure in the microſcope, and tranſparently, ſeem to have been generated out of the water.

Firſt, I ſay, that their tranſparency which they diſcovered in the microſcope, is an argument, becauſe I believe there is no tranſparent body in the world, that has not been reduced to that conſtitution, by being ſome ways or other made fluid; nor can I indeed imagine how there ſhould be any. All bodies, made tranſparent by art, muſt be reduced into that form firſt; and therefore 'tis not unlikely but that nature may take the ſame courſe; but this, as only probable, I ſhall not inſiſt on. Next, I ſay, that the figures of divers of them in the microſcope diſcover the ſame things; for I have ſeen multitudes of them curiouſly wrought and figured like cryſtals or diamonds; and I cannot imagine by what other inſtrument nature ſhould thus cut them, ſave by cryſtallizing them out of a liquid or fluid body; and that way we find her to work in the formation of all thoſe curious, regular figures of ſalts, and the vitriols, as I may call them, of metals and divers other bodies, of which chymiſtry affords many inſtances. Sea-ſalt and ſal-gem cryſtallizes into cubes or four-ſided parallelopipeds; nitre into triangular and hexangular priſms; alum into octahedrons; vitriols into various kinds of figures, according to the various kinds of metals diſſolved, and the various menſtruums diſſolving them; tartars alſo, and candyings of vegetables are figured into their various regular ſhapes from the ſame method and principle: and in truth, in the formation of any body out of this mineral kingdom, whoſe origin we are able to examin, we may find that nature firſt reduces the bodies to be wrought into a liquid or ſoft ſubſtance, and afterwards forms and ſhapes it into this or that figure. But this argument drawn from the ſand, found in all running ſtreams, I ſhall not inſiſt on, becauſe ſome imagine it to be only waſht off from the land and ſhores the rivers paſſed over, and perhaps much of it may: but yet that ſand may be made of clear water, my ſecond argument will manifeſt, and that is this:

That

That 'tis a ufual experiment in the making of falt in the falterns, by the boiling up, or evaporating away the frefher part of the fea-water, to collect great quantities of fand at each corner of the boiler; which after it has been well wafh'd with frefh water, is, in all particulars, a perfect fand; and yet the water is fo ordered before it is put into the boiler, that nothing of fand or dregs can enter with it, the brine being firft fuffered to ftand a good while and fettle in a very large fat, fo that all the fand and dregs may fink to the bottom; after which the clearer water at the top is drawn off, and fuffered to run into the boiler. 'Tis not impoffible perhaps, but that fubftance which made the fand, might be diffolved in the water, and afterwards by evaporation coagulated; which if fo, makes not at all againft, but rather argues ftrongly for my fourth propofition.

But that the other folution is fomething more probable, namely, that 'tis made out of the very fubftance of the water itfelf, this third argument will make probable; and that is, that any water, of what kind foever, though never fo clear and infipid, may, by frequent diftillations, be all of it perfectly tranfmuted into a white infipid calx, not again diffolvable in water, and in nothing differing from the fubftance of ftone. This I have been affured by an eminent phyfician, who has divers times made trial of it with the fame fuccefs. If therefore the whole body of any water may, by fo eafy an operation, in fo very fhort a time, be tranfmuted into a ftony fubftance, what may not nature do, that can take her own time, and knows beſt how to make uſe of her own principles?

But, fourthly, we have many inſtances, by which we are aſſured that nature really does change water into ſtone, both by forming in a little time, conſiderable ſtones out of the diſtilling drops of water ſoaking through the roofs of caves and ſubterraneous vaults, of which we have very many inſtances here in *England*; as, to name one for all, at the *Peak* in *Derbyſhire*, the pendulous cones of this petrified ſubſtance directly point at, and oftentimes meet and reſt upon the riſing ſpires, generated by the drops of water trickling through the roof, as I mentioned before.

And, fifthly, there are divers other waters which we need not ſeek after in caves that have a petrifying virtue, and incruſtate all the channel they paſs through, and the ſubſtances ſoak'd in them, with ſtone; theſe are ſo common almoſt in all places, that I need not inſtance in any; only I cannot paſs by one, taken notice of by *Kircher*, being obſervations made by himſelf, and it has in it two circumſtances very conſiderable; the firſt is, that vegetables ſhould grow ſo plentifully in a very hot water; the ſecond, that only ſuch herbs as grew in it, and not ſuch as were ſteeped in it, will perfectly, after drying, be turned into ſtone, of which I ſhall have occaſion to make more uſe. I ſhall give the hiſtory in his own words [a]. "*Hæc experientia didici in itinere meo Hetruſco, in quo prope Roncolanum, Senenſis territorij oppidum* (a town near *Siena* in *Tuſcany*) *duos fontes calidos obſervavi, quorum aqua per*

[a] Mund. ſubterr. lib. v. ſect. 2. parag. 7.

cana-

canales ad molares rotas vertendas ducebatur. In hisce canalibus cyperus, junci, ranunculus similesque herbæ tanta adolescebant fœcunditate, ut quotannis eas, ne aquæ motum inturbarent, extirpare oporteret: extirpatas vero projectasque in vicinum locum, herbas omnes in lapidem conversas, non sine admiratione spectavi. Cujus rei causam cum a molitoribus quærerem; responderunt aquas istiusmodi hujus virtutis esse, ut quæcunque inter canales, aut ipsâ aquâ excreverint herbæ, mox ac extirpatæ fuerint, lapidescant; quæcunque vero extra aquam, in campis patentibus excreverint herbæ, istas extirpatas nunquam lapidescere. I pass by his reasons and explications, because I think them very little to the purpose: but the observations themselves are very considerable, and serve for the explaining of several phænomena I have observed in petrified bodies, as I shall endeavour hereafter to shew, as in corals, both white and red, and the several rarities of them; in corallines also, and petrified mushrooms, of each of which I have examined a very great variety. But this only by the by.

Sixthly, therefore 'tis observable that these petrifying waters are for the most part very clear and limpid; so that to the sight not to be distinguishable from other water, but only by the effects; and therefore, by the newly mentioned observations of *Kircher*, we find that vegetables, which upon drying, turned into stone, whilst green and growing, flourished and spread faster than others: so that the petrifying substance past through the finest and closest pores of the living vegetables, and therefore must certainly be very intimately mixed with the water that could not be ſeparated by ſo fine and curious ſtrainers.

But, ſeventhly, to confirm this propoſition yet farther, there are found in ſeveral parts of the earth ſuch waters as will be entirely converted into ſtone. Of this kind there are ſeveral hiſtories in the newly mentioned book, which I paſs over, and ſhall only take notice of one for all, and that is an account ſent to the *Roman* college of Jeſuits from the maſters, ſurveyors and clerks of the *Hungarian* mines, in anſwer to ſome queries propounded to them. To the query concerning the properties and metallick experiments about mineral waters, they anſwer [b]. *Datur in fodinis aquæ genus quod in figuram ſaccabaro haud abſimilem degenerat, videlicet in lapillos albos.*

And again, from another prefect of the imperial mines in *Hungary*, in anſwer to the ſame query, we have this account [c]. *Reperitur quoque aqua quædam alba quæ in lapidem durum abit. Si vero hæc aqua ante ſuam coagulationem mineram cupream tranſiverit, tunc generatur ex ea lapis qui Malochites vocatur: quando vero aqua illa perfluit cupream mineram continentem argentum, fiet ex ea pulcher lapis ceruleus, ſimilis Turcoidi. Hæc aqua autem nullibi frequentius reperitur, quam in mineris lapidibus ſiliceis copioſis, et cuprum cum argento continentibus.* Whence I am apt to think, and I have many obſervations and arguments to prove my conjecture,

That, eighthly, all kinds of talk and ſpar, moſt ores and marchaſites, *Alumen plumeum* and *Abeſtus*, fluors, cryſtals, *Corniſh* diamonds, amethyſts, and

[b] *Kircher.* mund. ſubterr. p. 183. [c] *Id.* p. 185.

divers

divers other figured mineral bodies, may be generated from their cryſtalliſation, or coagulation, out of ſome mineral waters.

And to make it yet more probable, I could, in the ninth place, add divers experiments, by which ſeveral of theſe concretes may be in a ſhort time made artificially by ſeveral chymical operations, which would very much illuſtrate the former doctrine. But I hope what I have mentioned may ſuffice to make the fourth propoſition probable, that waters of divers kinds may be turned in time into ſtone, without being reducible again to water, by any art yet commonly known, which being granted, my

Fifth propoſition will follow of conſequence; namely, that divers other fluid ſubſtances, have, after long continuance of reſt, ſettled and congealed into much more hard and permanent ſubſtances: for if water itſelf may be ſo changed and metamorphoſed, which ſeems the fartheſt removed from the nature of a ſolid body, certainly thoſe which are nearer to that nature, and are mixed with ſuch waters, will more eaſily be coagulated. I ſhall not therefore any farther inſiſt on the proof of this, than only to mention two particulars, and that becauſe we have almoſt every where ſo many inſtances and experiments; the firſt is of *Pliny* [d] in theſe words: *Verum et ipſius terræ ſunt alia ſegmenta. Quis enim ſatis miretur, peſſimam ejus partem, ideoque pulverem appellatam in Puteolanis collibus, opponi maris fluctibus, merſumque protinus fieri lapidem unum inexpugnabilem undis, et fortiorem quotidie, utique ſi Cumano*

[d] Lib. xxxv. cap. 13.

miſ-

misceatur cæmento. Eadem est terræ natura in Cyzicena regione: sed ibi non pulvis, verum ipsa terra qualibet magnitudine excisa et demersa in mare, lapidea extrahitur. Hoc idem circa Cassandriam produnt fieri: et in fonte Gnidio dulci intra octo menses terram lapidescere. Ab Oropo quidem Aulidem usque quicquid terræ attingitur mari, mutatur in saxa, &c. To the end of the chapter he goes on to relate divers places where earths, *&c.* are turned into stones. And in another place [e] he tells us, "*Nitrariæ egregiæ Ægyptijs: nam circa Naucratim et Memphim tantum solebant esse, circa Memphim deteriores: lapidescit ibi in acervis: multique sunt tumuli ea de causa saxei: faciunt ex his vasa,* &c.

The second is an observation of my own, which I have often taken notice of, and lately examined very diligently; which will much confirm these histories of *Pliny*, and this my present hypothesis; and that is a part of the observation which I made on the western shore of the Isle of *Wight*. I observed a cliff of a pretty height, which, by the constant washing of the water at the bottom of it, is continually, especially after frosts and great rains, foundering and tumbling down into the sea underneath it. Along the shore underneath this cliff, are a great number of rocks and large stones confusedly placed, some covered, others quite out of the water; all which rocks I found to be compounded of sand and clay, and shells, and such kind of stones as the shore was covered with. Examining the hardness of some that lay as far into the water, as the low-water-mark, I found them to be altogether as hard, if not much harder than

[e] Lib. xxxi. cap. 10.

Port-

Portland or *Purbeck* ſtone. Others of them, that lay not ſo far into the ſea, I found much ſofter, as having in probability not been ſo long expos'd to the viciſſitudes of the tides: others of them I found ſo very ſoft, that I could eaſily with my foot cruſh them, and make impreſſions into them, and could thruſt a walking ſtick I had in my hand, a great depth into them. Others that had been but newly founder'd down, were yet more ſoft, as having been ſcarce waſhed by the ſalt water: All theſe were perfectly of the ſame ſubſtance with the cliff, from whence they had manifeſtly tumbled, and conſiſted of layers of ſhells, ſand, clay, gravel, earth, &c. and from all the circumſtances I could examine, I do judge them to have been the parts of the neighbouring cliff tumbled down, and rowl'd and waſhed by degrees into the ſea; and by the petrifying power of the ſalt water, converted into perfect hard compacted ſtones. I have likewiſe ſince obſerved the ſame phænomena on other ſhores: and I doubt not but any inquiſitive naturaliſt may find infinite of the like inſtances all along the coaſt of *England*, and other countries where there are ſuch kind of foundering cliffs. I ſhall not now mention the great quantities of toothed ſpar, which I obſerved to be cryſtallized upon the ſides of theſe rocks, which ſeem'd to have been nothing elſe but the meer cryſtallizing or ſhooting of ſome kind of water, which was preſs'd or aroſe out of theſe coagulating ſtones: for the hiſtory of theſe kinds of figured ſtones belong more properly to another diſcourſe; namely of the natural geometrical figures obſervable in ores, minerals, ſpars, talk, &c. of which elſewhere.

One

One inſtance more I cannot omit, as being the moſt obſervable of any I have yet heard of; and that is Dr. *Caſtle*'s relation of a certain place at *Alpſly* in *Bedfordſhire*, where there is a corner of a certain field, that doth perfectly turn wood and divers other ſubſtances in a very ſhort time into ſtone, as hard as a flint or agate. A piece of this kind I ſaw, affirm'd to have been there buried, which the perſon that had buried it, had ſhot ſmall ſhots of lead into The whole ſubſtance of the wood, bark and pith, together with the leaden ſhot itſelf, was perfectly turned to a ſtone as hard as any agate, and yet retained its perfect ſhape and form; and the lead remained round, and in its place, but much harder than any iron.

But to ſpend no more time on the proof of that of which we have almoſt every where inſtances, divers of which I have already mentioned, I ſhall proceed to the ſixth propoſition; which is, that a great part of the ſurface of the earth hath been ſince the creation transformed, and made of another nature: that is, many parts which have been ſea are now land, and others that have been land are now ſea; many of the mountains have been vales, and the vales mountains, *&c.*

For the proving of which propoſition I ſhall not need to produce any other arguments, beſides the repeating what I find ſet down by divers natural hiſtorians concerning the prodigious effects that have been produced by earthquakes, on the ſuperficial parts of the earth; becauſe they ſeem to me to have been the chief efficients which have tranſported the petrified bodies, ſhells, woods, animal

ſubſtances, &c. and left them in ſome parts of the earth, as are no other ways likely to have been the places wherein ſuch ſubſtances ſhould be produced; they being uſually either raiſed a great way above the level ſurface of the earth, on the tops of hills, or elſe buried a great way beneath that ſurface in the lower vallies: for who can imagine that oiſters, muſcles, periwinkles, and the like ſhell-fiſh ſhould ever have had their habitation on the tops of the mountain *Caucaſus?* Which is by divers of our geographers accounted as high in its perpendicular altitude, as any mountain in the yet known world; and yet *Olearius* affords us a very conſiderable hiſtory to this purpoſe, of his own obſervation, which I ſhall hereafter have occaſion to relate, and examine more particularly. Or, to come a little nearer home, who would imagine that oiſters, *Echini*, and ſome other ſhell-fiſh, ſhould heretofore have lived at the top of the *Alpes*, *Apennine*, and *Pyrenean* mountains, all which abound with great ſtore of ſeveral ſorts of ſhells; nay, yet nearer, at the tops of ſome of the higheſt in *Cornwal* and *Devonſhire*, where I have been informed by perſons whoſe teſtimony I cannot in the leaſt ſuſpect, that they have taken up divers, and ſeen great quantities of them? And to come yet nearer, who can imagine oiſters to have lived on the tops of ſome hills near *Banſtead Downs* in *Surry?* Where there have been time out of mind, and are ſtill to this day found divers ſhells of oiſters, both on the uppermoſt ſurface of the earth, and buried likewiſe under the ſurface of the earth, as I was lately informed by ſeveral very worthy perſons living near
thoſe

thoſe places, and as I myſelf had the opportunity to obſerve and collect.

Of the Effects of EARTHQUAKES.

TO proceed then to the effects of earthquakes, we find in hiſtory four ſorts or genus's to have been performed by them.

The firſt, is the raiſing of the ſuperficial parts of the earth above their former level: and under this head there are four ſpecies. The firſt is the raiſing of a conſiderable part of a country, which before lay level with the ſea, and making it lye many feet, nay, ſometimes many fathoms above its former height. A ſecond, is the raiſing of a conſiderable part of the bottom of the ſea, and making it lye above the ſurface of the water, by which means divers iſlands have been generated and produced. A third ſpecies is the raiſing of very conſiderable mountains out of a plain and level country. And a fourth ſpecies is the raiſing of the parts of the earth, by the throwing on of a great acceſs of new earth, and ſo burying the former ſurface under a covering of new earth many fathoms thick.

A ſecond ſort of effects performed by earthquakes, is the depreſſion or ſinking of the parts of the earth's ſurface below the former level. Under this head are alſo compriſed four diſtinct ſpecies, which are directly contrary to the four laſt named.

The firſt, is a ſucking of ſome part of the ſurface of the earth, lying a good way within the land,

and converting it into a lake of almoſt an unmeaſurable depth.

The ſecond, is the ſinking of a conſiderable part of plain land, near the ſea, below its former level, and ſo ſuffering the ſea to come in, and overflow it, being laid lower than the ſurface of the next adjacent ſea.

A third, is the ſinking of the parts of the bottom of the ſea much lower, and creating therein vaſt *vorages* and *abyſſes.*

A fourth, is the making bare, or uncovering of divers parts of the earth, which were before a good way below the ſurface; and this either by ſuddenly throwing away theſe upper parts, by ſome ſubterraneous motion, or elſe by waſhing them away by ſome kind of eruption of waters, from unuſual places, vomited out by ſome earthquake.

A third ſort of effects produced by earthquakes, are the ſubverſions, converſions, and tranſpoſitions of the parts of the earth.

A fourth ſort of effects, are liquefaction, baking, calcining, petrifaction, transformation, ſublimation, diſtillation, &c.

The firſt therefore of the effects of earthquakes, which I but now named, was, that divers parts of the ſurface of the earth, which lay below, or level with the ſea, have been raiſed a good height above that level, by earthquakes. Of this *Pliny* gives us ſeveral inſtances. [f] *Eadem naſcentium cauſa terrarum eſt, cum idem ille ſpiritus attollendo potens ſolo non valuit erumpere. Naſcuntur enim nec fluminum tantum invectu, ſicut Echinades inſulæ ab Acheloo*

[f] Hiſt. nat. lib. ii. cap. 85.

amne

amne congestæ; majorque pars Ægypti a Nilo, in quam a Pharo insula noctis et diei cursum fuisse Homero credimus: sed et recessu maris sicut eidem Circeis. Quod accidisse et in Ambraciæ portu decem millium passuum intervallo, et Atheniensium quinque millium ad Piræeum memoratur: et Ephesi, ubi quondam ædem Dianæ alluebat. Herodoto quidem si credimus, mare fuit supra Memphim usque ad Æthiopum montes: itemque a planis Arabiæ. Mare et circa Ilium, et tota Teuthrania, quaque campos intulerit Mæander.

And *Sandys* also, in his travels through *Italy*, and the parts of the *Levant*, gives this instance[g], speaking of the new mountain, which was produced in the kingdom of *Naples*, in the year 1538, says, "The lake *Lucrinus* extended formerly to *Avernus*, "and so unto *Gaurus*, two other lakes; but is now "no other than a little sedgy plash, choaked up "by the horrible and astonishing eruption of a new "mountain, whereof, as oft as I think, I am apt "to credit whatsoever is wonderful. For who in "*Italy* knows not, or who elsewhere will believe, "that a mountain should arise, partly out of a "lake, and partly out of the sea in one day and "a night, to such a height, as to contend in al"titude with the high mountains adjoining!

"In the year of our Lord 1538 on the 29th of "*September*, when for certain days foregoing, the "country thereabouts was so vext with perpetual "earthquakes, as no one house was left so intire, as "not to expect immediate ruin, after that the sea "had retired 200 paces from the shore, leaving "abundance of fish, with springs of fresh water

[g] P. 277.

"rising

" rising at the bottom, this mountain visibly as-
" cended about the second hour of the night, *&c.*"
And again [h], speaking of the same place, " The
" sea was accustomed, when urged with storms,
" to flow in through the lake *Lucrinus*, driving
" fishes in with it; but now, not only that pas-
" sage, but a part of *Avernus* itself is choak'd up
" by the mountain."

In which histories I take notice only of these two particulars at present. First, that that part of the land which lies between *Lucrinus* and the sea, that was oft-times before overflowed by the sea; since this earthquake has been so far raised, as that now such effects are no longer to be found. To confirm the rising of which the more, the other circumstance of the sea's departing from the shore 200 paces does much contribute. But, not to insist on this, Mr. *Childrey*, in his *Britannia Baconica*, a book very useful in its kind, being a collection of all the natural history of the islands of *Great Britain*, to be met with in *Cambden* or *Speed*, and some other historians, together with such of his own as he had opportunity to observe, relates to us many considerable passages to this purpose. In his history of *Norfolk*, he says, " that near St. *Be-*
" *net*'s in the *Holm*, are perfect cockles and peri-
" winkles sometimes digged up out of the earth,
" which makes some think it was formerly over-
" flowed by the sea." The fenny grounds also of *Lincolnshire* and *Cheshire*, seem to have proceeded from the rising of the ground; and those in *Anglesy*, where lopp'd trees are now digged up with

[h] Page 281.

the perfect ſtrokes of the ax remaining on them, ſeem to have been firſt ſunk under water, then overturned and buried in their own earth, and afterwards the whole earth ſeems to have been raiſed again to its former height.

Linſchoten gives us a relation of the like effects that happened in the *Tercera's*. The relation, as I find it epitomiz'd by *Purchas*[i] is this: "In *July* "*anno* 1591, there happened an earthquake in the "iſland of St. *Michael*, which lieth from *Tercera* "ſouth about 28 miles, an iſland 20 miles long, "and full of towns, which continued from *July* "26 to *Auguſt* 12, in which time none durſt ſtay "within his houſe, but fled into the fields, faſting "and praying with great ſorrow, for that many "of their houſes fell down, and a town called "*Villa Franca*, was almoſt razed to the ground, "all the cloyſters and houſes ſhaken to the earth, "and therein people ſlain. *The land in ſome places* "*roſe up*, and the clifts removed from one place "to another, and ſome hills were defaced and "made even with the ground. The earthquake "was ſo ſtrong, that the ſhips which lay in the "road, and in the ſea, ſhaked as if the world "would have turned round. There ſprung alſo "a fountain out of the earth, from whence, for "the ſpace of four days, there flowed a moſt clear "water, and after that it ceaſed. At the ſame time "they heard ſuch thunder and noiſe under the "earth, as if all the devils had been aſſembled "together at that place, wherewith many died for "fear. The iſland of *Tercera* ſhook four times

[i] *Pilgrim.* part. iv. p. 1677.

" together, ſo that it ſeemed to turn about; but
" there happened no other misfortune unto it.
" Earthquakes are common in thoſe iſlands: for
" about 20 years paſt there happened another
" earthquake, when a high hill that lieth by the
" ſame town *Villa Franca*, fell half down, and co-
" vered all the town with earth, and killed many
" men." I have tranſcribed here, once for all, the whole relation, becauſe there are many other conſiderable circumſtances in it beſides the riſing of the earth, which I ſhall have occaſion to refer to, under others of the heads or propoſitions to be proved, and therefore ſhall not need repetition. Two other relations I find collected by *Purchas*, confirming this, and ſeveral of the other propoſitions: the one is that of *Dithmar Blefken* in the hiſtory of *Iſland*[k]. " On the 29th of *November* a-
" bout midnight, in the ſea, there appeared a
" flame near *Hecla*, which gave light to the whole
" iſland: an hour after the whole iſland trembled,
" as it would have been moved out of the place:
" after the earthquake followed a horrible crack,
" that if all war-like ordnance had been diſcharg-
" ed, it had been nothing to this terror. It was
" known afterwards that *the ſea went back two*
" *leagues in that place, which remained dry*."

A ſecond hiſtory *Purchas* has collected out of the hiſtory of *Joſeph Acoſta* of the *Weſt Indies*[l]; o-mitting for the preſent divers other circumſtances he takes notice of, I ſhall only mention that of the receding of the ſea." " Upon the coaſt of *Chili*, I
" remember not well in what year, there was ſo ter-

[k] *Id.* part iii. p. 648.

[l] Part iii. p. 940.

" rible an earthquake, as it overturned whole
" mountains, and thereby ſtopt the courſe of ri-
" vers, which it converted into lakes. It beat
" down towns, and ſlew a great number of peo-
" ple, cauſing the ſea to leave her place ſome
" leagues, ſo as the ſhips remained on dry ground,
" far from the ordinary road, *&c.*" An example ſomewhat like this happen'd lately in the *Eaſt Indies*, as I was lately informed by a letter from thence. The thing in ſhort was this: at a place about ſeven days journey from *Ducca*, the earth trembled about 32 days; and the ſequel was, that it raiſed the bottom of a lake, ſo as to drive out all the water and fiſh upon the land, ſo that a place which was formerly a lake is now dry ground. This was written from *Ballaſore*, *Jan.* 6, 1665.

The ſecond ſpecies of effects of earthquakes, is the raiſing a conſiderable part of the bottom of the ſea, and making it lie above the ſurface of the water, by which means ſeveral iſlands have been generated. Of this *Pliny* gives us ſeveral inſtances[m]. *Naſcuntur et alio modo terræ* (having in the preceding chapter ſpoken of the ſhores riſing above the water, or the waters deceding from the ſhore) *ac repente in aliquo mari emergunt, veluti paria ſecum faciente natura, quæque hauſerit hiatus, alio loco reddente. Claræ jampridem inſulæ, Delos et Rhodos, memoriæ produnt enatæ. Poſtea minores, ultra Melon Anaphe* (of which *Strabo* makes mention[n].) *Inter Lemnum et Helleſpontem Nea. Inter Lebedum et Teon, Alone: inter Cycladas, Olympiadis* CXXXV. *anno* 4to, *Thera et Theraſia. Inter eaſdem poſt ann.* CXXX *Hiera:*

[m] Lib. ii. cap. 86, 87. [n] Lib. x.

et

et ab ea duobus ſtadijs poſt ann. cx *in noſtro ævo, M. Junio Syllano, L. Balbo COSS. ad* VIII *Idus Julias, Thia.* Two of which hiſtories are alſo confirmed by *Seneca* [o], where explicating the effects of earthquakes by the commixture of fire and water, he ſays, *Theren et Theraſiam et hanc noſtræ ætatis inſulam, ſpectantibus nobis in Ægæo mari enatam quis dubitat quin in lucem ſpiritus vexerit. Sandys* ſpeaking of the *Iolian* iſlands, ſaith, " Of thoſe there were only ſeven, now there are eleven in number, " which heretofore all flamed; now only *Vulcano* " and *Strombylo*, two of that number, do burn." *Vulcano* is ſaid firſt of all to have appeared above water about the time that *Scipio Africanus* died. But we have much later inſtances to confirm this our aſſertion: for about 28 years ſince, an iſland was made among the *Azores* by an eruption of fire, of which divers have related the ſtory. But *Kircher* [p], from the relation of the Jeſuits, has added the moſt particular one. Having ſpoken of the exceeding height of the *Pike* of *Teneriffe* in the *Canaries*, and of the eruptions of fire in it, and the hot ſprings round about it, he adds, that in the *Azores* alſo there are found places having almoſt the ſame properties. The *Pico de Fayal de Santo Gregorio*, being almoſt of equal height, and St. *Michael*'s iſland having had heretofore ſeveral *Vulcans*, and having been troubled with many earthquakes, and very notably about 38 years ſince, wherein all the iſland was ſo terribly ſhaken, that the utter ruin and ſubverſion of the whole was feared. The hiſtory of which, in ſhort, is this; that " *June*

[o] Quæſt. nat. lib. vi. [p] Mund. ſubterr.

" 26, 1638, the whole iſland began to be ſhaken
" with earthquakes for eight days, ſo that the in-
" habitants left cities, caſtles and houſes, and dwelt
" in the fields, but eſpecially thoſe of a place call-
" ed *Vargen*, where the motion was more violent.
" After which earthquake this prodigy followed.
" At a place of the ſea, where fiſhermen uſed to
" fiſh in ſummer, becauſe of the great abundance
" of fiſh there caught, called *La Femera*, about
" ſix miles from *Pico Delle Carmerine*, upon the firſt
" ſunday in *July*, a ſubterraneous fire, notwith-
" ſtanding the weight and depth of the ſea in that
" place, which was 120 foot, as the fiſhermen had
" often before that found by ſounding, and the
" multitude of waters, which one would have
" thought ſufficient to have quenched the fire: a
" ſubterraneous fire, I ſay, broke out with a moſt
" inexpreſſible violence, carrying up into the clouds
" with it water, ſand, earth, ſtones, and other
" vaſt bulks of bodies; which to the ſad ſpecta-
" tors, at a diſtance, appeared like flocks of wool
" or cotton, and falling back on the ſurface of
" the water, look'd like froth. The ſpace of this
" eruption was about as big as a ſpace of land,
" that might well be ſown by two buſhels of grain.
" By great providence the wind blew from the
" land; otherwiſe the whole iſland would, in all
" probability, have periſhed by the mercileſs rage
" of theſe devouring flames. Such vaſt bulks of
" ſtone were thrown up into the air, about the height,
" to ſeeming, of three pikes length, that one would
" rather think them mountains than rocks: and
" which added further to this dreadful ſight, was,
" that

"that these mountains returning again, often met
"with others ascending or being thrown up, and
"were thereby dasht into a thousand pieces; di-
"vers of which pieces being afterwards taken up
"and bruised, easily turned into a black shining
"sand. Out of the great multitude and variety
"of these vast rejected bodies, and the immense
"heaps of rocks and stones, after a while was
"formed a new island out of the main ocean,
"which at first was not above five furlongs over;
"but after a while, by daily accesses of new mat-
"ter, it increased after fourteen days to an island
"of five miles over. From this eruption, so great
"a quantity of fish was destroyed and thrown up-
"on the next adjacent island, that eight of the
"biggest *Indian* galeons would not be sufficient to
"contain them; which the inhabitants fearing,
"lest the stink of them might create a plague, for
"eighteen miles round collected and buried in
"deep pits. The stink of the brimstone was
"plainly smelt at 24 miles distance." But we
have an instance more of the generation of an island out of the bottom of the sea, by an eruption;
which because it happened very lately, namely in
1650, and near an island in the *Archipelago*, which
Pliny relates to have been heretofore after the same
manner produc'd, I shall in short relate, as it is
more largely recorded by *Kircher* [q] from the mouth
of Father *Franciscus Riccardus*, a Jesuit, who was
at the same time in the adjoining island, and an
eye witness of all the phænomena.

[q] Mund. subterran.

" From the 24th of *September* to the 9th of *Oc-*
" *tober* 1650, the iſland of *Santerinum*, formerly
" called by *Pliny*, *Thera*, was dreadfully ſhaken
" with earthquakes, ſo that the inhabitants expect-
" ed nothing but utter ruin; and were yet more
" amazed by a horrid eruption of fire out of the
" bottom of the ſea, about four miles to the eaſt-
" ward of the iſland: before which the water of
" the place was raiſed above 30 cubits perpendi-
" cularly. (I ſuppoſe he means as to appearance
" from the iſland, otherwiſe 'tis but very little)
" which wave ſpreading itſelf round every way,
" overturned every thing it met, deſtroying ſhips
" and galleys in the harbour of *Candie*, which was
" 80 miles diſtant. The eruption filled the air
" with aſhes and horrible ſulphureous ſtinks, and
" dreadful lightnings and thunder ſucceeded. All
" things in the iſland were covered with a yellow
" ſulphureous cruſt, and the people almoſt blind-
" ed as well as choaked. Multitudes of pumice
" and other ſtones were thrown up, and carried
" as far as *Conſtantinople*, and to places at a very
" great diſtance. The force of this eruption was
" greateſt the two firſt months, when all the
" neighbouring ſea ſeemed to boil, and the *Vul-*
" *can* continually vomited up fire balls. Upon
" the turning of the wind, great miſchief was
" done in the iſland of *Santerinum*; many beaſts
" and birds were killed: and on the 29th of *Octo-*
" *ber* and the 4th of *November*, about 50 men
" were killed by it. The other four months it
" laſted, tho' much abated of its former fierce-
" neſs, yet it ſtill caſt up ſtone, and ſeemed to endeavour

"deavour the making of a new iſland; which tho' "it do not yet perfectly appear above water, yet "'tis covered but eight foot by the water; and the "bubbling of the water ſeems to beſpeak another "eruption, that may in time finiſh natures birth." And though our natural hiſtorians have been very ſcarce in the world, and conſequently ſuch hiſtories are very few; yet there has been no age wherein ſuch hiſtorians have lived, but has afforded them an example of ſuch effects of earthquakes. And I doubt not, but had the world been always furniſhed with ſuch hiſtorians as had been inquiſitive and knowing, we ſhould have found not only *Thera* or *Santerinum*, and *Vulcano*, and *Delos*, and that in the *Azores*, and one lately in the *Canaries*, but a very great part of the iſlands of the whole world, to have been raiſed out of the ſea, or ſeparated from the land, by earthquakes: for which opinion I ſhall afterwards relate ſeveral obſervations both of my own and others, which ſeem to afford probable arguments.

But to proceed to the third kind or ſpecies of effects produced by earthquakes, which is the raiſing very conſiderable mountains out of plains. Of this I ſhall add a few inſtances; but none more notable, than that of the new mountain near *Naples*, of which I ſaid ſomewhat before out of *Sandys*'s travels. In the year 1538, *Sept.* 29th, this mountain viſibly aſcended about the ſecond hour of the night, with a hideous roaring, horribly vomiting ſtones, and ſuch ſtore of cinders as overwhelmed all the buildings thereabouts, and the ſalubrious baths of *Tripergula*, for ſo many ages ce-

 lebrated,

celebrated, conſuming all the vines to aſhes, and killing birds and beaſts, and frightning away all the inhabitants, who fled naked and defiled through the dark: and has advanced its top a mile above the baſis: the ſtones of it are ſo light and porey that they will not ſink when thrown into the ſea. This new mountain, when new raiſed, had a number of iſſues, ſome of them ſmoking, ſome flaming, others diſgorging rivulets of hot water, keeping within a terrible rumbling; and many periſhed that ventur'd into the hollowneſs above. But that hollow on the top is at preſent an orchard, and the mountain throughout bereft of its terrors. "It is re-
"ported, ſays *Childrey* [r], that by the ſea ſide, not
"far from *Axbridge* in *Somerſetſhire*, within theſe
"50 years, a parcel of land ſwell'd up like a hill,
"but on a ſudden clave aſunder, and fell down
"into the earth; and in the place of it remains a
"great pool." Our *Engliſh* chronicles ſay, at *Oxenhal*, in the biſhoprick of *Durham*, on *Chriſtmas* day 1179, the ground heav'd up aloft like a tower, and continued all the day unmovable, till evening, and then fell with a horrible noiſe, ſinking into the earth, and leaving three deep pits called hell-kettles. *Vorenius* [ſ] tells us of a new mountain likewiſe raiſed in *Java*, in the year 1586, with the like effects of thoſe I formerly named of the new mountain; firſt ſhaking the earth, then heaving up, and throwing up into the air, the upper parts of the earth, afterwards the rock and inner parts, then fiery coals and cinders, overwhelming the circumjacent fields and towns, and hilling above

[r] Britan Baconic. [ſ] Geogr.

10,000

10,000 men, and burning what was not overwhelmed. I have not time to reckon up the multitude of inſtances I have met with in authors; ſuch as *Ætna* in *Sicily*, *Veſuvius* in *Italy*, one in *Croatia*, near the city *Valonia*, the *Pike* in *Teneriffe*, and the *Pike* in the *Azores*, *Hecla*, *Helga*, and another in *Iſland:* the mount *Gonnapi* in one of the iſlands of *Banda*, which made an horrid eruption at the ſame time with that in *Java*. The mount *Balavane* in *Sumatra:* others in the *Molucca* iſlands, in *China*, *Japan*, and the *Philippines*, and in ſome of the *Maurician* iſlands, and ſeveral other parts of the *Eaſt Indies*. In the *Weſt Indies* alſo we have multitudes of examples; ſeveral in *Nicaragua*, and all along the ledge of mountains in *Peru* and *Chili*, and in *New Spain* and *Mexico:* in the iſlands of *Papoys*, diſcovered by *La Mair* joining to the ſouth continent in *Mar Del Zur:* all which are ſo many ſhining torches to direct us in the ſearch after this truth. There are many other inſtances of mountains, that have but lately, as it were, left to burn, and are covered with wood and grown fruitful. So the new mountain I formerly mentioned, has an orchard growing where the fire at firſt flamed. Another in the iſland *Quimada*, near the river *Plata* in *Braſil:* the iſlands alſo of St. *Helena*, and *Aſcenſion*, diſcovered by the great plenty of cinders, and the faſhions of the hills, to have formerly contained *Vulcanos*, and probably were at firſt made by ſome ſubterraneous eruption, as indeed moſt of thoſe iſlands in the main ocean, ſuch as the *Canaries*, and the *Azores*, and the *Eaſt Indian*, and the *Caribbee* iſlands, and divers others ſeem to have been.

been. A paſſage to make this aſſertion ſomewhat the more probable, I have met with in *Linſchoten*'s deſcription of the iſle of *Tercera*, which, as *Purchas* has epitomized[t], I have here added. "The land "is very high, and, as it ſeemeth, hollow; for "that as they paſs over an hill or ſtone, the ground "ſoundeth under them as if it were a cellar: ſo "that it ſeems in divers places to have holes un-"der the earth, whereby it is much ſubject to "earthquakes, as alſo all the other iſlands are; "for there it is a common thing: and all thoſe "iſlands, for the moſt part, have had mines of "brimſtone; for that in many places of *Tercera* "and St. *Michael*, the ſmoke and ſavour of brim-"ſtone doth ſtill iſſue out of the ground, and the "country round about is all ſinged and burnt. "Alſo there are places wherein there are wells, "the water whereof is ſo hot that it will boil an "egg, as if it were over a fire." Beſides which, the ſhape of the hills, and ſeveral other circumſtances mentioned by *Linſchoten*, do make it probable that thoſe have been all *Vulcano's*.

But to proceed to the fourth ſpecies of the effects of earthquakes under this head; and that is, the raiſing of the parts of the earth, by the throwing on a great acceſs of new earth: of this I have already given many inſtances in the newly mentioned hiſtories of eruptions, where I mentioned the overwhelming of fields, towns and woods. I ſhall only add one inſtance or two more to confirm this head, and then proceed. The firſt is that mentioned by *Olaus Wormius*[u], where he gives an ac-

[t] Pilgr. part iv. p. 1670. [u] Muſæi. lib. i. ſect. i. chap. 5.

count

count of an extraordinary earthquake in *Iceland*, which filled the air with duſt, earth, and cinders, and overwhelmed towns, fields, and even ſhips a good way diſtant at ſea; and which ſent forth its fumes with ſuch violence and plenty, as covered all the decks and ſails of ſhips lying on the coaſt of *Norway*, ſome hundred leagues diſtant. And to make this of *Wormius* the more probable, I have now by me a paper of duſt, which was rained out of the air upon a ſhip lying at *Algier* upon the coaſt of *Barbary*, upon a great eruption of *Veſuvius* in the year 1600. But what is beyond all, is the late eruption of *Mongibell* or *Ætna*.

And to confirm this propoſition yet further, I cannot paſs by a very remarkable rain of earth and aſhes, that happened in *Peru*, *anno* 1600, mentioned by *Garcelaſſo de la Vega*, one of the offspring of the *Inca's* of *Peru*, in his hiſtory of *America*. The epitome of which by *Purchas*[w], is this. "I might "add the great earthquakes *anno* 1600 in *Peru* at "*Arequepa*, the raining of ſand, as alſo of aſhes a-"bout 20 days from a *Vulcano* breaking forth; "the aſhes falling in places above a yard thick, "in ſome places more than two, and where leaſt, "above a quarter of a yard, which buried the corn "grounds of maize and wheat, and the boughs "of trees were broken and fruitleſs, and the cat-"tle great and ſmall died for want of paſture: for "the ſand which rained covered the fields thirty "leagues one way, and above forty leagues ano-"ther way. Round about *Arequepa* they found "their kine dead by 500 together in ſeveral herds,

[w] Pilg. part iv. p. 1476.

"and

" and whole flocks of ſheep, and herds of goats
" and ſwine buried. Houſes fell with the weight
" of the ſand, and others coſt much induſtry to
" ſave them. Mighty thunders and lightning
" were heard and ſeen 30 leagues about *Arequepa.*
" It was ſo dark whilſt thoſe ſhowers laſted, that
" at mid-day they burned candles to ſee to do
" buſineſs."—I could add divers other inſtances
to confirm this propoſition, but theſe may at preſent ſuffice.

But this is but one way by which divers things have been buried: there is another way which I can only at preſent mention, and muſt refer the probation and proſecution to ſome other occaſion; and that is, that very many of the lower ſuperficial parts of the earth, have been, and are continually covered and buried by the acceſs of matter, tumbled and waſhed down by exceſſes of wind and rain, and by the continual ſweepings of rivers, and ſtreams of water. Under this head I ſhall ſhew ſeveral places and countries in the world, that are nothing elſe but the productions of theſe cauſes. To this purpoſe *Peter de la Valle* [x] gives ſome obſervations which he made in *Egypt*, " Of the former ſeven
" mouths of *Nile*, there are only four left, and of
" thoſe but two navigable; the reſt are either fill-
" ed, or run no more, or are ſmall ſtreams not
" taken notice of, or only torrents in the time of
" great rains: but I could learn nothing of them,
" becauſe the great expence of the ancients for
" cleanſing the ditches, has been intermitted for
" ſeveral hundreds of years." He is likewiſe of

[x] Letter xith dated from *Grand Cairo, Jan.* 25, 1616.

opinion,

opinion, with *Herodotus*, that the *Delta*, and all the lower *Egypt*, where the *Greeks* navigated in his time, was in the firſt ages of the world made by the ſand and mud of *Nile*.

All which hiſtories and particulars do manifeſtly enough evince, that there have been in very many parts of the world conſiderable mutations of the ſuperficial parts, ſince the beginning; and that therefore thoſe places where theſe petrified bodies are found, though they now ſeem never ſo much foreign, and differing from the likely native places of ſuch animated bodies, may notwithſtanding heretofore have been in ſuch another kind of condition, as was moſt ſuitable to the breeding and nouriſhment of them, which I ſhall yet further manifeſt, by comparing the other effects produced by earthquakes; ſuch as the ſinking, and burying, and tranſpoſing, and overturning of the ſuperficial parts of the earth.

Another ſort of effects, is the ſinking of the ſuperficial parts of the earth, and placing them below their former poſition, both in reſpect of ſome parts newly raiſed, and in reſpect of ſome other adjacent parts not diſplaced. And this ſeems to be cauſed by the ſubſiding or ſinking thoſe parts into ſuch caverns, as by the ſtrength of the eruption paſſing below, before it breaks out, are made underneath. And if the parts of the earth underneath are ſo looſe or obnoxious to the force of the fire, as to be diſlodged; unleſs the remaining parts be very ſtrong, and conſtitute a very firm ſtony arch, the earth does very eaſily tumble into the holes and hollows made by the fire. Now it cannot be imagined

gined but that all thoſe vaſt congeries of earth, which I have already mentioned to have been thrown up, and to create new iſlands and new mountains, and the like, muſt leave vaſt caverns below them, to be filled, either with the parts of the earth that hang immediately over them, or with the ſea, or other ſubterraneous waters, if the roofs of theſe cavities be ſtrong enough to ſuſtain the earth above them from ſinking. And ſome ſuch power as theſe ſubterraneous fires, ſeems to me to have been the cauſe of the ſtrange poſitions and intermixture of the veins of ores and minerals in the bowels of the mountains, where, for the moſt part, they are now found; and even of bringing thoſe ſubſtances ſo near the ſurface of the earth, which from the conſideration of very many circumſtances, ſeem to me to be naturally ſituated at a much greater depth below within the bowels of this globe. And hence may be rendred a reaſon of the figures of thoſe minerals, and other of thoſe ſubſtances mixed with them, and of the compounding and blending of ſeveral of theſe ſubſtances together, whereby ſome of them are very ſtrangely united and alter'd. But this I mention only by the by, and ſhall not inſiſt on it, belonging more properly to another head. To proceed then, under this general head are compriſed ſeveral kinds of effects, differing only according to the parts of the earth they have been wrought upon.

The firſt is, the ſinking of ſeveral inland parts, which were before eminent, and laying them much lower into vales. Sometimes, the ſinking of a part of the earth to a very great depth, and leav-

ing

ing behind, inſtead of a firm ground, a lake of ſalt or ſea-water. Of theſe we have ſeveral inſtances in natural hiſtorians. And, to paſs by many others, I ſhall only mention ſuch as have lately happened. Of this kind Mr. *Childrey* in his *Britannia Baconica*, has collected ſeveral inſtances, two out of our *Engliſh* chronicles: his relations are theſe[y]. "*Auguſt* the 4th, 1585, after a very vi-
" olent ſtorm of thunder and rain, at *Mottingham*
" in *Kent*, eight miles from *London*, the ground ſud-
" denly began to ſink; and three great elms grow-
" ing upon it, were carried ſo deep into the earth,
" that no part of them could any more be ſeen.
" The hole left (ſaith the ſtory) is in compaſs 80
" yards about, and a line of 50 fathoms, plumm'd
" into it finds no bottom." Alſo, "*December* 18,
" 1596, a mile and half from *Weſtram* ſouthward
" (which is not many miles from *Mottingham*) a
" part of an hedge of aſhes, 12 perches long, was
" ſunk ſix foot and an half deep; the next morn-
" ing 15 foot more; and the third morning 80
" foot more at leaſt, and ſo daily." (And preſently afterwards he ſays) " Moreover in one
" part of the plain field there is a great hole made
" by ſinking of the earth, to the depth of 30 foot
" at leaſt, being in breadth in ſome places, two
" perches over, and in length five or ſix perches.
" There are ſundry other ſinkings in divers other
" places; one of 60 foot, another of 47, and an-
" other 34 foot; by means of which confuſion it
" is come to paſs, that where the higheſt hills

[y] Page 62.

" were, there be the loweſt dales, and the loweſt
" dales are become the higheſt grounds, *&c.*"

And again [z], he gives an inſtance upon his own knowledge, much to the ſame purpoſe, which lately happened. " *July* the 8th 1657, about three
" of the clock, in the pariſh of *Bickly*, was heard
" a very great noiſe like thunder afar off; which
" was much wondered at, becauſe the ſky was
" clear, and no appearance of a cloud. Shortly
" after (ſaith the author of this relation) a neigh-
" bour came to me, and told me, I ſhould ſee a
" very ſtrange thing if I would go with him. So
" coming into a field, called the *Lay-Field*, we
" found a very great bank of earth, which had
" many tall oaks growing on it, quite ſunk into
" the ground, trees and all. At firſt we durſt
" not go near it, becauſe the earth, for near 20
" yards about, was exceedingly much rent, and
" ſeemed ready to fall: but ſince that time myſelf
" and ſome others have ventur'd to ſee the bot-
" tom, I mean to go to the brink, ſo as to diſcern
" the viſible bottom, which is water, and con-
" ceived to be about 30 yards from us; under
" which is ſunk all the earth about it, for 16 yards
" round at leaſt, three tall oaks, a very tall awber,
" and certain other ſmall trees, and not a ſprig of
" them to be ſeen above water. Four or five oaks
" more are expected to fall every moment, and
" a great quantity of land is like to fall, indeed
" never ceaſing more or leſs; and when any con-
" ſiderable clod falls, it is much like the report
" of a cannon. We can diſcern the ground hol-

[z] Page 131.

" low

" low above the water a great depth; but how far
" hollow, or how deep, is not to be found out by
" man. Some of the water (as I have been told)
" drawn out of this pit with a bucket, was found
" to be as ſalt as ſea water, *&c.*"

" A conſiderable circumſtance alſo to confirm this propoſition, is a paſſage in that hiſtory I have mentioned out of *Linſchoten*, of the iſland of *Tercera*, where he ſays, *and ſome of the hills were defaced and made even with the ground.*

Kircher tells us a very remarkable hiſtory of the ſinking of a town, and the land about it, and the generation of a lake inſtead of it[a]. *Contigit hac eadem horâ res æternâ ac immortali memoriâ digna*, &c. " At this very time happened a thing wor-
" thy never to be forgotten, to wit the ſubverſion
" of the moſt famous town called St. *Euphemia:*
" 'twas ſituated at the ſide of the bay under the ju-
" riſdiction of the knights of *Malta*. When
" therefore we had come to *Lopiz*, almoſt dead
" from the violent ſhaking of the earth, and lying
" proſtrate on the ground, at laſt the *Paroxyſm* of
" nature remitting, caſting our eyes towards the
" neighbouring places, we ſaw the forementioned
" town encompaſſed with a great, wonderful, and
" unuſual cloud, which was ſeen by us three times,
" eſpecially at three o'clock in the afternnon, the
" heavens being clear. This cloud being by de-
" grees diſſipated, we look'd for the town, but
" found it not; a ſtinking lake, to our wonder,
" appearing in the place of it. We ſought for
" ſome perſon or other, to give us ſome certain
" account of this unuſual event; but could not

 " find

[a] Mund. ſubterr. præfat. cap. 2.

"find one to tell any news of this dreadful acci-
"dent and great deſtruction, *&c.* We proſecuting
"our journey, and paſſing by *Nicaſtro*, *Amantea*,
"*Paula*, *Belvedere*, found nothing for 200 miles,
"but the remaining carcaſſes of cities and caſtles,
"and horrid deſtructions; the men lying in the
"open fields, and, as it were, dead and withered
"through fear and terror."

To this purpoſe give me leave to ſubjoin an extract of a letter, ſent from *Balaſore* in the *Eaſt Indies*, *Jan.* 6, 1665. "The ſame ſtar appeared in
"our horizon about the ſame time 'twas ſeen with
"you. The effects have in part been felt here
"by unſeaſonable weather, great mortalities a-
"mong the natives, *Engliſh*, and others. We have
"had ſeveral earthquakes, unuſual here, which with
"hideous noiſes, have in ſeveral places broke out,
"and ſwallowed up houſes and towns. But about
"ſeven days journey from *Ducca*, where were at
"that time three or four *Dutch*, they and the na-
"tives relate, that in the market place, the earth
"trembled about 32 days and nights, without in-
"termiſſion. At the latter end, in the market
"place, the ground turned round as duſt in a
"whirlwind, and ſo continued ſeveral days and
"nights, and ſwallowed up ſeveral men who were
"ſpectators, who ſunk and turned round with the
"earth, as in a quagmire. At laſt the earth
"worked and caſt up a great fiſh, bigger than
"hath been ſeen in this country, which the peo-
"ple caught: but the concluſion of all was, that
"the earth ſunk with 300 houſes, and all the men,
"where now appears a large lake ſome fathoms

"deep. About a mile from this town was a lake
"full of fish, which in these 32 days of the earth-
"quake cast up all her fish on dry land, where
"might have been gathered many, which had
"run out of the water upon dry land, and there
"died: but when the other great lake appeared,
"this former dried up, and is now firm land."

To the same purpose also we have several other instances, some later, and some nearer home. "Near *Darlington*, (says *Childrey* [b], speaking of
"the rarities of the bishoprick of *Durham*) are
"three pits, whose waters are warm, (hot says
"*Cambden*) wonderful deep, called hell-kettles.
"These are thought to come of an earthquake,
"that happened *anno* 1179. For on *Christmas*
"day, say our chronicles, at *Oxenhall*, which is
"this place, the ground heaved up aloft like a
"tower, and so continued all that day, as it were
"immoveable, till evening, and then fell in with a
"very horrible noise, and the earth swallow'd it up,
"and made in the same place three deep pits."
The same, in the section of *Brecknock*, says, "Two
"miles east of *Brecknock*, is a meer, called *Llin-*
"*savathan*, which, as the people dwelling there
"say, was once a city; but the city was swallow-
"ed up by an earthquake, and this water or lake
"succeeded in the place; the lake is encompassed
"with steep high hills, *&c.*"

"Near *Falkirk*, says *Lithgow*, remain the ruins
"and marks of a town, *&c.* swallow'd up into
"the earth by an earthquake, and the void place
"is filled with water." *Pliny* also records a like

[b] Præfat. Mund. subterr. cap. 2.

inſtance[c], *Mox in his montem Epopon*, &c. " Pre-
" ſently the mountain *Epopon* (when ſuddenly a
" flame had ſhone out of it) was levell'd with the
" plain, and in the ſame plain a town was ſwal-
" lowed up into the deep, and by another motion
" of the earth became a lake. And in another
" place the mountain being tumbled down, the
" iſland *Prochyta* aroſe, *&c.*"

The *Dead Sea* alſo in *Paleſtine*, was the production of a moſt terrible earthquake, and a fire ſent from heaven. For, methinks, the relation of the ſad cataſtrophe of thoſe four cities, *Sodom*, *Gomorrha*, *Zeboim*, *Adma*, mentioned in ſcripture, ſeem ſomewhat like that I have newly related out of *Kircher* of St. *Euphemia*. There are a multitude of other inſtances which I could bring on this head, of the ſinking of mountains and hills into plains, and all theſe into lakes, of which *Pliny* gives ſeveral inſtances[d]. The *Pico* in the *Molucca's*, accounted of equal height with that of *Teneriffe*, was, by a late earthquake, quite ſwallowed into the earth, and left a lake in its place. *Veſuvius* and *Strongylus*, are by late earthquakes reduced to almoſt half their former height. Many of thoſe vaſt mountains of the *Andes* in *Chili*, were by an earthquake *ann.* 1646, quite ſwallowed up and loſt, as *Kircher* relates. I could add many hiſtories of the fatal cataſtrophe's of many towns, and other places of note; but theſe, I hope, may ſuffice to ſhew this kind alſo of mutation in the ſuperficial parts of the earth, to be effected by earthquakes.

[c] Britann. Baconic. [d] Hiſt. nat. lib. ii. cap. 88.

Nor

Nor do earthquakes only ſink mountains and in-land parts; but ſuch parts alſo as are near to, e-qual with, and under the ſurface of the ſea. Of this we have inſtances near home, of *Winchelſea* and of the *Goodwin Lands*, and of the towns in *Freezland*, that have been about 400 years ſince ſwallowed up by the ſea; and nothing but towers and the *Goodwin Sands*, are now to be found of them. The like happened to ſeveral parts of *Scotland*, as *Hector Boethius* relates. *Linſchoten*, in his hiſtory of the *Weſt Indies*, relates, among many other hiſtories this conſiderable paſſage. "Since, "in the year 1586, in the month of *July*, fell "another earthquake in the city of *Kings*, the "which, as the vice-roy did write, had run 170 "leagues along the coaſt, and athwart to the *Sierra* "50 leagues. It ruined a great part of the city. "It caus'd the like trouble and motion of the "ſea, as it had done at *Chili*, which happened "preſently after the earthquake; ſo as they might "ſee the ſea to fly furiouſly out of her bounds, and "to run near two leagues into the land, riſing a-"bove fourteen fathom.....It covered all the plain, "ſo as the ditches were filled, and pieces of wood "that were here, ſwam in the water." There are multitudes of inſtances of the like effects in ſe-veral other parts of the world, which have been wrought by earthquakes, which may be found in natural hiſtorians; which, for brevity's ſake, I o-mit, they ſerving only to prove a propoſition, which I ſuppoſe will be granted by any that have either ſeen or heard of the effects of earthquakes.

Now, though I find a general deficiency in natural hiſtorians, of inſtances to prove, that the ſubmarine parts have likewiſe ſuffered the like effects of ſinking; they lying out of view, and ſo cannot without ſome trouble and diligence be obſerved; yet if we conſider from how great a depth theſe eruptions proceed, and how little diſtinction they make between mountains and plains, as to the weight of removing, we may eaſily believe, that the bottom of the ſea is as ſubject to theſe mutations, as the parts of the land. And ſince, by the former relations, we have many inſtances of the raiſing of the bottom of the ſea, 'tis very probable, that what quantity of matter is thrown to and raiſed in one place, is ſunk, and falls into that cavity left by another. An iſland cannot be raiſed in one place, without leaving an abyſs in another. And I do not doubt, but there have been as many earthquakes in the parts of the earth under the ocean, as there have been in the parts of the dry land: but being for the moſt part till of late, unfrequented by mankind, and even now but very thinly, 'tis almoſt a thouſand to one, that what happen are never ſeen; and a hundred to one, if they have been ſeen, whether they be recorded: for how few writers are there of natural hiſtory?—There is ſomewhat of probability in the ſtory related by *Plato*, in his *Timæus*, of the iſland *Atlantis*, in the *Atlantic* ocean, which, he ſays, was ſwallowed up by an earthquake, into the ſea. And 'tis not unlikely, but that moſt of thoſe iſlands that are now appearing, have been either thrown up out of the ſea by eruptions, ſuch as the *Canaries*, *Azores*,

res, St. *Helena*, &c. which the form of them, and the *Vulcano's* in them, and the cinders and pumice ſtones found about them, and the frequent earthquakes they are troubled with, and the remaining hills of extinguiſhed *Vulcano's*, do all ſtrongly argue for: or elſe, that they are, ſome of them at leaſt, ſome relicks of that great iſland which is now not to be found; and yet we have no records hereof. That there is as great inequality in the depth of the ſea, as there is in the height of the land, the obſervations of ſeamen, experimented by their ſounding-lines, do ſufficiently inform us: for hills, we have deep holes; and for mountains and pikes, abyſſes and malſtrooms: and that theſe muſt have, in all ages, been filling with parts of the earth, tumbled by the motion of the waters, and rowling to the loweſt place, is very probable; and ſo they would in time have been filled up, had not earthquakes, by their eruptions and tumblings, created new irregularities. And therefore that there are ſtill ſuch places, is an argument, that there have been of later ages earthquakes in ſome of them. Of theſe I ſhall mention one or two inſtances, which I meet with in voyages, and the relations of travellers.

In the relation of the circumnavigation of Sir *Francis Drake*, ſpeaking of the ſtraights of *Magellane*, he ſays, page 35, "They ſaw an iſland with a very high *Vulcano*;" and the next page he ſays, "They had need to have carried nothing but anchors and cables, to find ground, the ſea was ſo very deep;" which depth is explained more expreſsly page 42, where it is ſaid, "Being driven from

" from our firſt place of anchoring, ſo unmea-
" ſurable was the depth that 500 fathom would
" fetch no ground." And in page 99, of the ſame relation, the author tells, how their ſhip ſtruck upon a rock, which, page 102, he ſays " At low
" water was but ſix foot under water, and juſt by
" it no bottom to be found, by reaſon of the
" great depth."

Mr. *Ricaut*, in a letter of his to the Royal Society, dated from *Conſtantinople*, *Nov.* 1667, ſays,
" That the water runs out of the *Euxine* ſea into
" the *Propontis* with a wonderful ſwiftneſs, which is
" more wonderful in regard of the depth of the
" *Boſphorus*, being in the channel 50 or 55 fathom
" water, and along the land in moſt places the
" ſhips may lie on ſhore with their heads, and yet
" have 20 fathom water at their ſterns."

Beſides theſe effects of raiſing and ſinking the parts of the earth, there is a third ſort, which is the tranſpoſing, converting, ſubverting, and jumbling the parts of the earth together; overthrowing mountains, and turning them upſide down; throwing the parts of the earth from one place to another; burying the ſuperficial parts, and raiſing the ſubterraneous. Of theſe kinds of changes, there are many inſtances in the former relations I have mentioned, and particularly that of *Linſchoten* of the earthquake in the *Terceras*, and that of *Joſephus Acoſta*, of the earthquake upon the coaſt of *Chili*. And there are a multitude of others I could here ſet down, but I ſhall only mention ſome of them.
" Soon after (ſays *Acoſta*, in the place before men-
" tioned, which was in the year 1582) happened
" that

" that earthquake of *Arequipa*, which in a manner " overthrew the whole city." And a little before in the ſame place, he tells of a terrible earthquake in *Guatimala*, in the year 1586, which overthrew almoſt all the city, and that the *Vulcan* for above ſix months together continually vomited a flood of fire from the top of it. And a little after the ſame author, in the ſame place, ſays " In the year " 1581, in *Cugiano*, a city of *Peru*, otherwiſe call- " ed the *Pear*, there happened a ſtrange accident " touching this ſubject; a village called *Angoango*, " (where many *Indians* dwelt that were ſorcerers " and idolaters) fell ſuddenly to ruin, ſo as a great " part thereof was raiſed up and carried away, and " many of the *Indians* ſmothered; and that which " ſeemed incredible, yet teſtified by men of credit, " the earth that was ruined and ſo beaten down, " did run and ſlide upon the land for the ſpace of " a league and a half, as if it had been water or " wax melted, ſo as it ſtopped and filled up a " lake, and remained ſo ſpread all over the whole " country."

Nor are there wanting examples of this kind even in this iſland. Mr. *Childrey* [e] has collected ſeveral out of *Cambden*; as that in *Herefordſhire*, where " In the year 1571, *Marcley Hill* in the eaſt " part of the ſhire, with a roaring noiſe, removed " itſelf from the place where it ſtood, and for three " days together travelled from its old ſeat. It " began firſt to take its journey *Feb.* 17, being " *Saturday*, at ſix of the clock at night, and by " ſeven the next morning it had gone 40 paces,

[e] Britann. Baconic.

" car-

" carrying with it ſheep in their cotes, hedge-
" rows, and trees, whereof ſome were overturn-
" ed; and ſome that ſtood upon the plain, are
" firmly growing upon the hill; thoſe that were
" eaſt were turned weſt, and thoſe in the weſt
" were ſet in the eaſt. In this remove it over-
" threw *Kinaſton* chappel, and turned two high
" ways near a 100 yards from their old paths:
" the ground that they removed was about 26 a-
" cres, which opening itſelf with rocks and all,
" bore the earth before it for 400 yards ſpace with-
" out any ſtay; leaving paſturages in the places
" of tillage, and the tillage overſpread with paſtu-
" rage. Laſtly, overwhelming its lower parts,
" it mounted to a hill of 12 fathoms high, and
" there reſted after three days travel."

" At *Hermitage* in *Dorſetſhire*, ſays *Stow* in his
" *Summary*, *Jan.* the 3d 1582; a piece of ground
" of three acres removed from its old place, and
" was carried over another cloſe, where alders and
" willows grew, the ſpace of 40 rods or perches,
" and ſtopped up the highway that led to *Carne*,
" a market town; and yet the hedges that it
" was incloſed with incloſe it ſtill, and the trees
" ſtand bolt upright, and the place where this
" ground was, is left like a great pit." And 'tis
not a little obſervable that at the ſame time that
theſe changes happened in *America*, the like alſo
happened in *England*, of which I ſhall hereafter
give divers other inſtances, and ſhall alſo deduce
corollaries, that may otherwiſe ſeem very ſtrange,
and yet I queſtion not to prove the truth of them.

Pliny

Pliny ſays [f] *Maximus terræ memoria mortalium extitit motus*, &c. "There happened once (which I found
" in the books of the *Tuſcan* learning) within the
" territories of *Modena*, *L. Marcius* and *S. Julius*
" being conſuls, a great wonder of the earth: for
" two hills encountered each other, charging one
" another with a great craſh, and retiring again,
" a great flame and ſmoke in the day-time iſſuing
" out from between them to the ſky, while a great
" many of the *Roman* knights, their friends and
" travellers, beheld it from the *Æmylian* road.
" With this conflict, and meeting together, all the
" country houſes were daſhed to pieces, many a-
" nimals that were between them periſhed. This
" happened a year before the *Social* war. I know
" not whether it were more pernicious to *Italy* than
" the civil wars. No leſs a wonder was that in
" our age, in the laſt year of *Nero*, as we have
" ſhewn in our acts, when meadows and olive trees,
" the publick way lying between them, went into
" contrary (*exchanged*) places, in the *Marrucine*
" territory, on the lands of *Vectius Marcellus*, a
" *Roman* knight, procurator under *Nero*."

There are many of the like inſtances to be met with in authors, of the placing parts perpendicular or inclining, which were before horizontal; of the turning of other parts upſide downwards, of throwing parts from place to place; of ſtopping the paſſage of rivers, and turning them another way; of ſwallowing ſome rivers, and producing others anew; of changing countries from barren to fruitful, and from fruitful to barren; of making iſlands

[f] Hiſt. nat. lib. ii. cap. 18.

join

join to the continent, and ſeparating parts of the continent into iſlands. There are other relations which mention the vaſt ſpaces of ground, that have been all at once ſhaken and overturned; ſome of 500 miles in length, and a 150 in breadth: of the communication of *Vulcans*, which are, as it were, the noſtrils, or conſtant breathing places of theſe monſters, tho' placed at a very great diſtance one from another, by ſubterraneous caverns. Other relations furniſh us with inſtances of the ſubſtances they vomit out; ſuch as pumice ſtones, and ſeveral other ſorts of calcined and melted ſtones, and rocks, aſhes, minerals, hot water, ſulphur, flame, ſmoke, and various other ſubſtances.

In others we find inſtances of liquefactions, vitrifications, calcinations, ſublimations, diſtillations, petrifactions, transformations, ſuffocations, and infective, or deadly ſteams deſtroying all things near them, which probably may be one cauſe of the ſcarcity of relations, where 'tis probable, there have been ſo very many effects wrought in the world, of this kind. But theſe I ſhall not inſiſt upon.

There is only one thing more, that I think pertinent to our preſent purpoſe, and that is the univerſality of this active principle: there is no country, almoſt in the world, but has been ſome time or other ſhaken by earthquakes, that has not ſuffered ſome, if not moſt part of theſe effects. *Seneca* ſays [g] *omnia ejuſdem ſortis ſunt*, &c. " All things " are ſubject to the ſame chance; though they are " not yet moved they are moveable; for we err,

[g] In præfat. lib. vi. quæſt. nat.

" if

" if we believe any part of the earth excuſed and
" freed from this hazard; all are ſubject to the
" ſame law; nothing is made by nature ſo fixt as
" to be immoveable; ſome ſink at one time, ſome
" at another. And as in great cities, now this
" houſe, now that houſe hangs tottering on props;
" ſo on the great face of the earth, now this part
" prevails, now that. *Tyre* formerly was remarka-
" ble for its deſtruction: *Aſia* loſt at once twelve
" cities. Whatever the power may be, the for-
" mer year *Achara* and *Macedonia* felt it, now
" *Campania*. Fate takes its rounds, and repeats
" what it had long before acted: it brings ſome
" things often on the ſtage, ſome ſeldom; but ſuf-
" fers nothing to remain abſolutely free and un-
" touch'd. Not we men only are brought forth
" ſhort liv'd, frail beings: cities, countries, ſhores,
" nay the ſea itſelf, are the ſlaves of fate. Why
" therefore do we flatter ourſelves that the gifts of
" fortune will ſtick by us, or that happineſs will
" obſerve any rule or meaſure; happineſs, the
" moſt fleeting of all human things! They that pro-
" miſe to themſelves all things fixt, ſurely never
" think that the ground we ſtand upon is itſelf un-
" fixt. Nor was that the frailty only of *Campa-*
" *nia* or *Achaia*; 'tis the ſame in all ſoils and coun-
" tries, to be looſely joined and compacted, but
" eaſily, and by many ways diſſolved; the whole
" remains, while each part changes and ſinks into
" ruin and alteration."

Thus we ſee all countries in the world are ſubject to theſe convulſions, but thoſe moſt of all,
that are moſt mountainous: ſuch are uſually, all
the

the ſea coaſts. Therefore *Pliny* ſays, that "The "*Alpes* and *Apennine* mountains have very often "been troubled with earthquakes; maritime places "are moſt ſhaken, nor do the mountainous eſcape, "for I have found that the *Alpes* and *Apennines* "tremble."

For moſt probably thoſe that are moſt mountainous, are moſt cavernous underneath; to countenance which opinion, I have taken notice in certain very high cliffs towards the ſea, where the hills ſeemed as it were cleft aſunder, the one half having been probably foundered and tumbled into the ſea, and the other, as it were remaining, that at the bottom, near the water, for almoſt the whole length, there were very many large caverns, which by ſeveral circumſtances, ſeemed to be made before the acceſs of the ſea thereunto, and not by the waſhing and beating of the waves againſt the bottom of the cliffs: for I obſerved in many of them, that the plates or layers, as I may ſo call thoſe parts between the clefts in rocks, and cliffs to lean contrary ways, and not to meet, as it were, at the top like the roof of a houſe; and others of them in other forms, as if they had been caverns left between many vaſt rocks tumbled confuſedly one upon another. And indeed I cannot imagine, but that under theſe mountains, iſlands, cliffs or lands, that have been much raiſed above their former level, there muſt be left vaſt caverns, whence all that matter was thrown, where probably may be the ſeat or place of the generation of thoſe prodigious powers. But this only by the by, for I intend not here to examine the cauſes of their beginnings, force, and powerful

powerful effects, nor of their remaining, ceasing, renewing, or the like: it being sufficient for my present purpose, to shew that they have been certainly observed to produce those extraordinary effects, from what cause soever they proceed: that they have been heretofore in many places where they have now ceased for many ages; and that they have lately happened in places where we have no history that does assure us they have been heretofore: that they have turned plains into mountains, and mountains into plains; seas into land, and land into seas: made rivers where there were none before, and swallowed up others that formerly were: made and destroyed lakes; made peninsulas islands, and islands peninsulas: vomited up islands in some places, and swallowed them down in others; overturned, tumbled and thrown from place to place cities, woods, hills, *&c.* covered, burnt, wasted, and changed the superficial parts in others; and many the like strange effects, which since the creation of the world, have wrought many very great changes on the superficial parts of the earth, and have been the great instruments or causes of placing shells, bones, plants, fishes, and the like, in those places, where, with much astonishment, we find them.

Concerning the vicissitudes that places are subject to in relation to earthquakes, I find a memorable passage sent by *Paul Ricaut*, Esq; now consul at *Smyrna*, dated *Nov.* 23, 1667. "*Constantinople* " is not now subject to earthquakes as reported in " former times, there having not happened in the " last seven years, in which I have been an inha-

 " bitant

"bitant there, above one of which I have been "ſenſible; but within the twenty days, in *Smyr-* "*na*, fell out an earthquake which dangerouſly "ſhook all the buildings, but did little or no "harm: the ſhips in the road, and others at an "anchor, about three leagues from hence, were "ſenſible of it. It is reported that this city hath "been already ſeven times devoured by earth- "quakes, and it is propheſied, that it ſhall be ſo "again as ſoon as the houſes reach the old caſtle "upon the top of the hill, on the ſide of which "remains the ruins of the old city, and the tomb "of St. *Polycarp*, St. *John*'s diſciple, ſtill pre- "ſerved by the *Greeks* in great veneration."

There is another cauſe which has been alſo a great inſtrument in promoting the alterations on the earth's ſurface, the motion of water; whether by its deſcent from on high, as in rivers, thro' the immediate fall of rain or ſnow, or by the melting of ſnow; or ſecondly, by the ſeas natural motions, as tides and currents; or thirdly, by its accidental motions from winds and ſtorms. Of each of theſe natural hiſtorians abound in inſtances. The former principle ſeems to be that which generates hills, holes, cliffs and caverns, and all irregularity and aſperity on the earth's ſurface; and this is what endeavours to reduce them to their priſtine evenneſs by waſhing down the tops of hills, and filling up the bottoms of pits, conſonant to all the other methods of nature in working with contrary principles; by which there is a kind of continual circulation. Water is raiſed in vapours by one quality, and precipitated in drops by another; the rivers run into the ſea, and the ſea ſupplies them. In the planets there is a projectile force which makes them endeavour to recede from the ſun, and an attractive power, which keeps them from receding. The air impregnates the ground in one place, and is impregnated by it in another; all things almoſt circulate and have their viciſſitudes: we have multitudes of inſtances of the waſting of the tops of hills, and of the filling and encreaſing of the plains or lower grounds; of rivers continually carrying along with them great quantities of ſand, mud, *&c.* from higher to lower places; of the ſeas waſhing cliffs away, and waſting the ſhores; of land-floods carrying away with them all things that ſtand in their way, and covering the lands with mud, levelling ridges and filling ditches. Tides and currents in the ſea act in all probability what floods and rivers do at land; and ſtorms effect that on the ſea-coaſt, that great land-floods do on the banks of rivers. *Egypt*, as lying very low, and yearly overflowed, is inlarged by the ſediment of the *Nile*; eſpecially towards thoſe parts where that river falls into the *Mediterranean*. The gulph of *Venice* is almoſt choak'd with the ſand of the *Po*. The mouth of the *Thames* is grown very ſhallow by the continual ſupply of ſand brought down with the ſtream. Moſt part of the cliffs which wall in this iſland, do yearly founder and tumble into the ſea. By theſe means many parts are covered and raiſed by mud and ſand, that lie almoſt level with the water, and others are diſcovered and laid open that for many ages have been hid.

Of this kind the Royal Society received a memorable account from the learned Dr. *Brown*, concerning a petrified bone of a prodigious bigneſs, diſcovered by the falling of ſome cliffs; the words of the relation are theſe, "This bone *(which is now in their repoſitory)* "was found laſt year 1666, "on the ſea ſhore, not far from *Winterton* in *Nor-* "*folk*, near the cliff after two great floods; ſome "thouſand loads of earth being torn away by the "rage of the ſea, as it often happens upon this "coaſt, where the cliffs conſiſt not of rock, but "of earth. That it came not out of the ſea may "be conjectured, becauſe it was found near the "cliff; and by the colour of it, for if out of the "ſea it would have been whiter. Upon the ſame "coaſt, but, as I take it, nearer *Haſborough*, di-"vers great bones are ſaid to have been found, "and I have ſeen a lower jaw containing teeth, "of a prodigious bigneſs, and ſomething petri-"fied. All that have been found on this coaſt, "were after the falling of ſome cliff: where the "outward cruſt is fallen off it clearly reſembles "the bones of whales, and great cetaceous ani-"mals, upon comparing it with the ſkull and "bones of a whale, which was caſt upon the coaſt "near *Wells*, and which I have by me, the weight "whereof is 55 pounds." To this may be added the *Chartham* news, or the diſcovery of the ſea-horſe, or *Hippopotamus*'s teeth printed in the *Philoſ. Tranſ.* N° 272, p. 882.

Nor are theſe changes now only, but they have, in all probability, been of as long ſtanding as the world. So 'tis probable there may have been ſe-

 veral

veral viciſſitudes of changes wrought on the ſame part of the earth: it may have been of an exact ſpherical form, with the reſt of the earths or planets, at the creation of the world, before the eternal command of the Almighty, that the waters under the heaven ſhould go to their place, which before covered the earth, ſo as that it was ἀόρατος καὶ ἀκατασκεύαστος καὶ σκότος ἐπάνω τῆς ἀβύσσου καὶ πνεῦμα θεοῦ ἐπεφέρετο ἐπάνω τοῦ ὕδατος, inviſible and incompleated, and the darkneſs of the deep was over it, (being all covered with a very thick ſhell of water which invironed it on every ſide, it being then, in all probability, created of an exact ſpherical figure; and ſo the waters, being of themſelves lighter than the earth, muſt equally ſpread themſelves over the whole ſurface of the earth) and where the breath of the Lord moved above or upon the ſurface of thoſe waters. It may, I ſay, in probability, have been then a part of the exact ſpherical ſurface of the earth, and upon the command that the waters under the air or atmoſphere (which ſeems to be denoted by στερέωμα or firmament; for the *Hebrew* word ſignifies an *expanſum*) ſhould be gathered together in one place, and that the dry land ſhould appear. It may have been by that extraordinary earthquake (whereby the hills and land were raiſed in one place, and that the pits or deeper places, whither the water was to recede, and be gathered together, to conſtitute the ſea were ſunk in another) raiſed perhaps to lie on the top of a hill, or in a plain, or ſunk into the bottom of the ſea, and by the waſhing of waters in motion, either carried to a lower place to cover ſome part of the vale, or elſe be covered with adventitious earth, brought down upon it from ſome higher place; which kinds of alterations were certainly very great by the flood of *Noah*, and ſeveral other floods we find recorded in heathen writers. If at leaſt there were not ſomewhat of an earthquake which might again ſink thoſe parts, which had been formerly raiſed to make the dry land appear, and raiſe the bottom of the ſea, which had been ſunk for the gathering together of the waters. (which opinion *Seneca* aſcribes to *Fabianus*) " *Ergò* (ſays he) *cum affuerit illa neceſſitas temporis, multà ſimul fata cauſas movent nec ſine concuſſione mundi tanta mutatio eſt, ut quidam putant, inter quos Fabianus eſt*. His deſcription of the manner and effects of a flood, is fine, and very ſuiting to my preſent hypotheſis. This part being thus covered with other earth, perhaps in the bottom of the ſea, may, by ſome ſubſequent earthquakes, have ſince been thrown up to the top of an hill, where thoſe parts, with which it was by the former means covered, may, in tract of time, by the fall and waſhing of waters, be again uncovered and laid open to the air, and all thoſe ſubſtances which had been buried for ſo many ages before, and which the devouring teeth of time had not conſumed, may be there expoſed to the light of the day.

There are yet two other cauſes of the mutation of the ſuperficial parts of the earth, which have wrought great changes in the world; and thoſe are, either the ſeas overflowing a country or place, forced by ſome violent ſtorms or hurricanes of wind, or through the overflowings of rivers by great

great falls of rain, or ſomething ſtopping their courſe. Of theſe we have many inſtances in voyages: and we have often here at *London* felt the effects of the wind driving in the tide with ſo great force as to have overflowed the banks and filled the ſtreets and cellars. "At *Chatmos* in *Lancaſhire*, ſays *Childrey* [h], is a low moſſy ground, "very large, a great part of which, according to "*Cambden*, not long ago, upon the brooks ſwelling high, was carried quite away with them, "whereby the rivers were corrupted, and a number of freſh fiſh periſhed. In which place now "lies a low vale watered with a little brook, "where trees have been digged up lying along, "which are ſuppoſed by ſome to have come thus. "The channel of the brooks being not ſcowered, "the brooks have riſen, and made all the land "mooriſh that lay lower than others, whereby the "roots of trees being looſened, by reaſon of the "boggineſs of the ground, or by the water finding a paſſage under ground, the trees have, either by their own weight, or by ſome ſtorm, "been blown down, and ſo ſunk into that ſoft "earth and been ſwallowed up: for 'tis obſervable "that trees are no where digged out of the earth "but where the ſoil is boggy; and even upon hills "ſuch mooriſh and moiſt grounds are commonly "found; the wood of ſuch trees burning very "bright, like touch-wood (which perhaps is by "reaſon of the bituminous earth in which they "have been ſo long) ſo as ſome take them for fir-"trees. Such mighty trees are often found in

[h] Britann. Baconic. p. 167, 168.

" *Holland*, which are thought to be undermined
" by the waves working into the ſhore, or by
" winds driven forwards and brought to thoſe
" lower places where they ſettled and ſunk."

Again[i], " The ſea has eaten a great part of
" the land away of the eaſtern ſhires. There are
" on the ſhore of this ſhire (*Cumberland*) trees diſ-
" covered by the winds ſometimes at low water,
" which are elſe covered over with ſand; and it is
" reported by the people dwelling thereabouts, that
" they dig up trees without boughs, out of the
" ground, in ſeveral places of the ſhire, and many
" trees are found and digged out of the earth of
" the iſle of *Man*."

Again[k], " In divers parts of the low grounds
" and champain fields of *Angleſea*, the inhabitants
" every day find and dig out of the earth, the bo-
" dies of huge trees with their roots, and fir-trees
" of a wonderful bigneſs and length."

Again[l], " At the ſame time that *Henry* II. made
" his abode in *Ireland*, were extraordinary violent
" and laſting ſtorms of wind and weather, ſo that
" the ſandy ſhore on the coaſts of *Pembrokeſhire*,
" was laid bare to the very hard ground, which
" had lain hid for many ages, and by further
" ſearch the people found great trunks of trees,
" which when they had digged up, they were ap-
" parently lopped, ſo that one might ſee the ſtrokes
" of the ax upon them, as if they had been given
" but the day before: the earth looked very black,
" and the wood of theſe trunks was altogether like
" ebony. At the firſt diſcovery made by theſe

i Britan. Baconic. p. 171. k Ib. p. 150. l Ib. p. 142, 143.

" ſtorms,

"ftorms, the trees we fpeak of lay fo thick, that
"the whole fhore feemed nothing but a lopped
"grove; whence may be gathered, that the fea
"hath overflowed much land on this coaft, as it
"has indeed many countries bordering upon the
"fea, which is to be imputed to the ignorance of
"the *Britons*, and other barbarous nations, who
"underftood not thofe ways to reprefs the fury of
"the fea, which we now do."

And again[m], "In the low places on the fouth
"fide of *Chefhire*, by the river *Wever*, trees are
"often found by digging under ground, which
"people think have lain buried there ever fince
"*Noah*'s flood. St. *Bennet*'s in the *Holm* hath
"fuch fenny and rotten ground (fays *Cambden*)
"that if a man cut up the roots or ftrings of trees,
"it flotes on the water. Hereabout alfo are coc-
"kles and periwinkles fometimes digged up out
"of the earth, which makes fome think, that it
"was formerly overflowed by the fea."

The *lignum foffile* which is found in *Italy*, of which we have a good·account given by *Francifco Stelluti*, from many circumftances of the hiftory, feems to me to have been firft buried by fome earthquakes, and afterwards to be varioufly metamorphofed by the fymptoms which ufually follow them, and which this place is much vexed with, as is indeed almoft all the country of *Italy*, for it emits hot fteams and fmoke proceeding from fubterraneous fires, which do there often fhift their places; burn the parts of fome of thofe trunks into black and brittle coals; melt a kind of ore into the pores of

[m] Ib. p. 129.

others;

others; petrify the ſubſtance of another ſort; bake the dirt and clayiſh ſubſtances which have ſoaked into the pores of a fourth ſort into a kind of brick; rot the parts of others, and convert them into a kind of dirt or muddy earth; and ſo act variouſly, and produce differing effects, upon thoſe buried ſubſtances, according to the nature of the earths, minerals, waters, ſalts, heats, ſmoaks, ſteams, and other active inſtruments caſually apply'd to the parts of the buried trunks, by the confuſion of the earthquakes, and by immediate application, and long continuance, and digeſtion, as I may call it, in this laboratory of nature, transformed into other ſubſtances, and exhibit all thoſe admirable phænomena mentioned by that author, whereby the buried bodies are transformed. Nor is it ſo much to be wondered at, that ſuch ſubſtances as vegetables, ſhould after many ages remain entire, and rather more ſubſtantially found than if they were newly cut down; ſince if we conſider the nature of decay and corruption in all kinds of animal and vegetable ſubſtances, we ſhall find that the chief cauſe of them is from the action of the fluid parts upon the ſolid, for the diſſolving of them: and whereſoever the internal fluid is either firſt changed, or altered by the admixture of ſome heterogeneous ſubſtance, ſo as to loſe that diſſolving property, as by the addition of ſalt, ſpirit of wine, *&c.* or by incorporating with it, and hardening it into a ſolid ſubſtance, as in petrifactions, *&c.* or ſecondly, exhaled by a gradual and gentle heat, and ſo the ſolid parts only left alone, and kept either dry, or filled with a fluid of an heterogeneous nature,

nature, ſuch as unctuous and ſpicy juices with watery ſubſtances. Or, thirdly, congealed and hardened, either by cold, or the peculiar nature of the juice itſelf; ſuch is freezing, and the hardening of the coralline plants, or ſubmarine vegetables, horns, gums, bones, hair, feathers, *&c.* Whereſoever, I ſay, bodies are by theſe means put into ſuch a conſtitution, that the parts act not, and continue in that ſtate, by being preſerved from adventitious moiſture, or ſoftening by homogeneous fluids; they are, as it were, perpetual, unleſs, by extraordinary heat, many of thoſe otherwiſe ſolid and unactive ſubſtances are made fluid by ſuch active diſſolvents; or unleſs they be immerſed in ſuch liquors or menſtruums as do of themſelves diſſolve and work on them; we ſhall not, I ſay, wonder at the laſtingneſs of theſe buried ſubſtances, if we conſider alſo the various juices with which ſeveral parts of the earth are furniſhed; unctuous, watery, ſtyptic, ſaline, petrifactive, corroſive, and what not. There are ſome juices of the earth which do, as it were, perpetuate them, by turning them into ſtone. Others do ſo deeply pierce and intimately mix with their parts, that they wholly, as it were, change the nature of their ſubſtances, and deſtroy that property of congruity which all bodies generated in the air and water ſeem to have, which are very apt to be diſſolved and corrupted by innate aerial and aqueous ſubſtances. Such are all kinds almoſt of oleaginous and ſulphureous bodies, and divers ſaline and mineral juices. Others indeed do not preſerve the very ſubſtance of thoſe vegetables, but by inſinuating

ating into their pores, and there, as it were, fixing, they retain and perpetuate the ſhape and figure, but corrupt and diſſolve the interpoſed part of the vegetable; of all which kinds I have ſeen ſome ſpecimens, as I have alſo of divers other ſubſtances, pickled, dried, candied, conſerved, preſerved, or mummified by nature. Where therefore the ſubſtances have happened to be buried with preſervative juices, they have withſtood the injury of time; but where thoſe juices have been wanting, there we find no footſteps of theſe monuments of antiquity.

But to return to what I was proſecuting; another cauſe which may make alterations on the ſurface of the earth, is any violent motions of the air, whereby the parts of the earth, in dry weather, are tranſported from place to place, in the form of duſt. Of this kind travellers tell us very ſtrange ſtories as to the removal of the ſands in the deſerts of *Arabia*, and other deſerts of *Africa*; and we have ſome inſtances of it here in *England*, to wit in *Norfolk* and *Devonſhire*, in the former of which there are often found natural mummies which have been buried alive by thoſe removing ſands, and by their dryneſs preſerved. But theſe greater and more ſudden removals of ſand and duſt are not ſo univerſal, and therefore not ſo much to my preſent purpoſe; though poſſibly they may have been more frequent heretofore, which the layers of ſands to be found in digging pits and wells ſeem to hint: but that which is moſt univerſal, is very ſlow, and almoſt imperceptible, namely the removing of the duſt from the higher parts, and ſettling thereof in the lower, by the wind or motion of the air.

I

I might mention alſo another cauſe of the tranſpoſition of the ſuperficial parts of the earth, and that is from the gradual ſubſiding or ſinking into the earth of the more heavy, and the ebullition, or reſpective riſing of the more light parts upwards. Hence we may obſerve, that many old and vaſt buildings and towers have ſunk into the earth: and the like we may judge of thoſe vaſt ſtones on *Saliſbury* plain, as we find conſtantly in almoſt all ſtone monuments placed in church-yards, and in all old churches, unleſs placed on a very high place, and founded on ſome rock. This cauſe may poſſibly have great influence where the earth is very ſoft, ſpungy, or boggy; and perhaps many of thoſe trees which are found in boggy grounds, may have been buried, by having been either felled, or blown down with wind, or waſhed down by ſome inundation, well impregnated with mineral juices, and ſo made heavier than the ſubjacent earth, and ſwallowed into it. Several of the former relations do indeed pretty well agree with this hypotheſis; and I am very apt to think that where the ſurface of the earth has not been much altered ſince the creation, if any ſuch there be; if it were ſearched into, it would be found that the lighteſt parts lie next the ſurface, and the heavier in the lower parts; which makes me imagine that the natural place of minerals is very deep under the ſurface of the earth, and poſſibly to be found under every ſtep of ground, were ſearch made to a ſufficient depth; and that the reaſon why we find them ſometimes near the ſurface, as in mountains, is not becauſe they were there generated, but becauſe they have been

been by ſome former ſubterraneous eruption, by which thoſe hills and mountains were made, thrown up towards the ſurface of the earth. And as gold is the heavieſt, ſo is it the ſcarceſt of all metals: Nor do I at all queſtion but that there may be other bodies or metals as much heavier than gold, as gold is heavier than common earth. To make theſe conjectures the more probable, ſee what Sir *Philiberto Vernatti* writes from *Batavia* in the *Eaſt Indies*, in anſwer to ſome queries ſent him by the Royal Society. "I have often felt earthquakes "here, but they do not continue long. In the year "1656, or 57, (I do not remember well the time) "*Batavia* was covered in one afternoon about two "of the clock, with a black duſt, which being "gathered together, was ſo ponderous that it ex-"ceeded the weight of gold. It is here thought "that it came out of a hill that burneth in *Suma-*"*tra*, near *Endrapeor*."

Theſe fiery eruptions in all probability come from a very great depth, and with a great violence; and poſſibly even that golden powder that is ſometimes thrown up, may have ſomewhat conduced to the cauſe of the violence of it. We know not what method nature may have to prepare an *aurum fulminans* of her own, great quantities of which, being any ways heated, and ſo fired, may have produced the powder. However, whether ſo or not, it is very well worth trial to examine, whether the flower which may be catch'd in a glaſs body, upon fulminating a quantity of ſuch powder gradually, by ſmall parcels, would, by being ordered as common gold, make again an *aurum fulminans*: or whe-

whether this fulmination, which is a kind of inflaming of the body of gold, does not make ſome very conſiderable alteration in the nature and texture of it.

But to proceed to the laſt argument to confirm the ſixth propoſition I at firſt undertook to prove; namely, that very many parts of the ſurface of the earth have been transformed, tranſpoſed, and many ways alter'd ſince the firſt creation of it. And that which to me ſeems the ſtrongeſt and moſt cogent argument of all, is this; that at the tops of ſome of the higheſt hills, and in the bottom of ſome of the deepeſt mines, in the midſt of mountains and quarries of ſtone, *&c.* divers bodies have been, and daily are found, which if we thoroughly examine, we ſhall find to be the real ſhells of fiſhes, which, for theſe following reaſons, we conclude to have been at firſt generated by the plaſtic faculty of the ſoul or life-principle of ſome animal, and not from the imaginary influence of the ſtars, or from any plaſtic faculty inherent in the earth itſelf ſo formed; the ſtreſs of which argument lies in theſe particulars.

I. That the bodies there found have exactly the form and matter, that is, are of the ſame kind of ſubſtance, for all its ſenſible properties, and have the ſame external and internal figure or ſhape with the ſhells of animals.

II. That it is contrary to all other acts of nature, that does nothing in vain, but always aims at an end, to make two bodies exactly of the ſame ſubſtance and figure, and one of them to be wholly uſeleſs, ufelefs, or at leaft, without any defign that we can with any plaufibility imagine.

III. Therefore, wherever nature works by peculiar forms and fubftances, we find that fhe always joins the body fo framed with fome other peculiar fubftance. Thus the fhells of animals, whilft they are forming are joined with the flefh of the animal to which they belong. Peculiar flowers, leaves and fruit are appropriated to peculiar roots, whereas thefe on the contrary are found mixed with all kind of fubftances, in ftones of all kinds, in all kinds of earth, fometimes expofed in the open air, without any coherence to any thing. This is, at leaft, an argument, that they were not generated in that pofture they are found; that very probably they have been heretofore diftinct and difunited from the bodies with which they are now mixt, and that they were not formed out of thefe very ftones or earth, as fome imagine, but derived their beings from fome preceding principle.

IV. Wherever elfe nature works by peculiar forms, we find her always to compleat that form, and not break off abruptly. But thefe fhells that are found in the middle of ftones, are moft of them broken, very few compleat, nay, I have feen many bruifed and flawed, and the parts at a pretty diftance one from another, which is an argument that they were not generated in the place where they were found, and in that pofture, but that they have been fometimes diftinct and diftant from thofe fubftances, and then only placed, broken and disfigured by chance, but had a preceding and more noble principle to which they owed their form, and

and by ſome hand of providence were caſt into ſuch places where they were filled with ſuch ſubſtances, as in tract of time have condenſed and hardened into ſtone. This, I think, any impartial examiner of theſe bodies will eaſily grant to be very probable, eſpecially if he takes notice of the circumſtances I have already mentioned. Now, if it be granted, that there have been preceding moulds, and that curiouſly figured ſtones do not owe their form to a plaſtic or forming principle inherent in their ſubſtances; why might not theſe be ſuppoſed ſhells, as well as other bodies of the ſame ſhape and ſubſtance, generated, none knows how, nor can imagine for what?

V. Further, if theſe be the apiſh tricks of nature, why does ſhe not imitate ſeveral other of her own works? Why do we not dig out of mines everlaſting vegetables, as graſs, for inſtance, or roſes, of the ſame ſubſtance, colour, ſmell *&c.* were it not that the ſhells of fiſhes are made of a kind of ſtony ſubſtance, which is not apt to corrupt and decay? Whereas plants and other animal ſubſtances, even bones, horns, teeth and claws, are more liable to the univerſal menſtruum of time. 'Tis probable therefore, that the fixedneſs of their ſubſtance has preſerved them in their priſtine form; and not that a new plaſtic principle has newly generated them. Beſides, why ſhould we not then doubt of all the ſhells taken up by the ſea ſhore, or out of the ſea, (if they had none when we found them) whether they ever had any fiſh in them or not? Why ſhould we not alſo here conceit a plaſtic faculty, diſtinct from that of the life principle

L of

of ſome animal? Is it becauſe this is more like a ſhell than the other? That, I am ſure cannot be. Is it becauſe it is more obvious how a ſhell ſhould be placed there than the other? If ſo, 'twould be as good reaſon to doubt, if an anchor ſhould be found at the top of a hill, as the poet affirms, or an urn or coins buried under ground, or in the bottom of a mine, whether it were ever an anchor or an urn, or a coined face, or made by the plaſtic faculty of the earth; than which, what could be more abſurd? And thoſe perſons that will needs be ſo over confident of their omniſcience of all that has been done in the world, or that could be, may, if they will vouchſafe, ſuffer themſelves to be aſked a queſtion, who informed them? Who told them where *England* was before the flood; nay even where it was before the *Roman* conqueſt, for about 4 or 5000 years, and perhaps much longer; much more where did they ever read or hear of what *changes* and tranſpoſitions there have been of the parts of it before that? What hiſtory informs us of the burying of thoſe trees in *Cheſhire* and *Angle-ſeu?* Who can tell when *Teneriffe* was made? And yet we find that moſt judicious men that have been there, and well conſidered the form and poſture of it, conclude it to have been at firſt that way produced. But I ſuppoſe the moſt confident will quickly, upon examination, find that there is a defect of natural hiſtory. If therefore we are left to conjecture, then that muſt certainly be the beſt that is backed with moſt reaſon; that clay, and ſand and common ſhells, can be changed and incorporated together into ſtones very hard. I have

 already

already given many inſtances, and can produce hundreds of others, but that I think it needleſs, that ſeveral parts of the bottom of the ſea have been thrown up into iſlands and mountains. I have alſo given divers inſtances, and thoſe, ſome of them, within the memory of man, where 'tis not in the leaſt to be doubted but that there may be found ſome ages hence ſeveral ſhells at the tops of thoſe hills there generated; and as little, that if quarries of ſtone ſhould be hereafter digged in thoſe places, there would be found ſhells incorporated with them; and were they not beholding to this inquiſitive and learned age for the hiſtory of that eruption, they might as much wonder how thoſe ſhells ſhould come there, and aſcribe them to a plaſtic faculty, or ſome imaginary influence, as plauſibly as ſome now do.

Now if all theſe bodies have been really ſuch ſhells of fiſhes as they moſt reſemble, and that they are found at the tops of the moſt conſiderable mountains in the world, as *Caucaſus*, the *Alpes*, the *Andes*, the *Apennine* and *Pyrenean* mountains, and that 'tis not very probable they were carried thither by mens hands, or by the deluge of *Noah*, or by any other more likely way than that of earthquakes; 'tis a very cogent argument that the ſuperficial parts of the earth have been very much changed ſince the beginning, that the tops of the mountains have been under water, and conſequently alſo that divers parts of the ſea's bottom have been heretofore mountains.

The ſeventh propoſition was, that 'tis very probable divers of theſe tranſpoſitions and metamor-

 phoſes

phoſes have been wrought here in *England*. Many of its hills have probably been heretofore under ſea. Of the latter of theſe I have given many inſtances already, and the firſt is probable from the great quantity of ſhells found in the moſt inland parts of the iſland; in hills, plains, bottoms of mines and middle of mountains and quarries. Of this kind are the infinite numbers in the *Portland*, *Purbeck*, *Burford* and *Northamptonſhire* ſtones: out of which I have often pickt muſcles, cockles, periwinkles, oiſters, ſcallops, *&c.*

'Tis improbable that either mens hands, or the general deluge, which laſted but a little while, ſhould bring them there: nothing more likely and ſufficient than an earthquake, which might heretofore raiſe theſe iſlands of *Great Britain* and *Ireland* out of the ſea, as it lately did thoſe in the *Canaries* and *Azores*, in the ſight of divers who are yet alive. Poſſibly *England* and *Ireland* might be raiſed by the ſame earthquake by which the *Atlantis*, if we will believe *Plato*, was ſunk.

Eighthly, that moſt of theſe mountains and inland places where theſe kind of petrified bodies and ſhells are found at preſent, or have been heretofore, were formerly under water; and that from the deſcending of the waters to ſome other place, by the tranſlation of the centre of gravity of the whole maſs, or rather by the eruption of ſome ſubterraneous fires, or earthquakes, great quantities of earth have been deſerted by the water, and laid bare and dry. That divers places have been ſo raiſed, has been already proved from many hiſtories; why then may not all of them have the ſame original?

original? There is no coin can ſo well inform an antiquary that ſuch and ſuch a place was once ſubject to ſuch a prince, as foſſil ſhells will certify a natural antiquary, that ſuch and ſuch places have been under water: and methinks providence ſeems to have deſigned theſe permanent ſhapes, as monuments and records to inſtruct ſucceeding ages of what paſs'd in preceding ones.

Ninthly, it ſeems probable, that the tops of the moſt conſiderable mountains in the world have been under water, and were raiſed to that height by ſome eruption; ſo that thoſe prodigious piles are nothing but the effects of ſome great earthquakes. This truth, 'tis likely, the poets have veiled under the feign'd ſtory of the giants, thoſe earth-born brothers, waging war with the gods, and heaping up mountains upon mountains; *Oſſa* and *Olympus* upon *Peleon*, and to hurl up great ſtones and fire againſt heaven, but that at laſt overcome by *Jupiter*'s thunder, they were buried under mountains, and the chiefeſt of them, *Typhæus* and *Enceladus* under *Sicily*, according to *Ovid* [n] and *Virgil* [o].

And as the poets had particular ſtories and giants for *Sicily* and *Ætna*, ſo had they alſo for other vulcano's and from the frequency of them in former ages about *Greece* and other parts of the *Mediterranean*: *Sophocles* calls them ὁ γηγενὴς στρατὸς γιγάντων, the earth-born army of the giants. And that nothing but earthquakes were meant by theſe giants, may be further collected from the place where they were ſaid to be bred, the *Phlegrean*

[n] Metamorph. lib. v. [o] Æneid. lib. iii.

fields in *Campania*, part of which, now called *Vulcan*'s court, is the vent of many ſubterraneous fires. Beſides, how well do their actions agree with the effects of earthquakes? For they are ſaid to throw up burning trees againſt heaven, and huge rocks, and vaſt hills, which, falling into the ſea, became iſlands, and mountains, lighting on the land. In a word, he that will read the deſcription of the moſt notable of them, *Typhæus*, and compare it with a natural deſcription of an earthquake, will eaſily explicate the ſeveral parts of the poets myſtical deſcriptions.

Though it be hard to prove this theory *poſitively*, thro' deficiency of natural hiſtory, yet if we conſider that the *Alps*, *Apennine* and *Pyrenean* hills, much the higheſt in *Europe*, have been infeſted with earthquakes, both formerly and lately, as we have ſeveral hiſtories that teſtify; and if other eruptions and earthquakes have raiſed mountains even out of the bottom of the ſea, and that the power of included fire is ſufficient to move and raiſe even a whole country all at once, for ſome hundreds of miles, as hiſtorians aſſure us; if we conſider all this I ſay, we may have reaſon to find it more than probable. And if to this we add the univerſal ſilence in hiſtory, of any part of *Europe*, nay of the whole world for almoſt 200 years after the flood, I think there will be much leſs ſcruple to grant that the many high mountains on whoſe tops are found ſuch numbers and varieties of true ſea ſhells, may have been heretofore raiſed up from under the ſea, and now are ſuſtained by the ſinking of other

other parts into the places from whence they were raised.

The tenth and last proposition is, that it seems not improbable but that the greatest part of the inequality of the earth's surface may have proceeded from the subversions and overturnings of some preceding earthquakes.

To prove this probability, I might repeat the argument, already urged; I could also instance in a multitude of other smaller effects of earthquakes, making the surface of the earth irregular, but they are so numerous and well known that I shall not insist on them. I might add also the universality of earthquakes, there being no part of the known world but we find to have been shaken by them. Thus much only I shall offer at present, that from what I have instanced about petrifactions, and hardening of several substances, it seems very likely that the earth in the beginning consisted for the most part of fluid substances, which by degrees have settled, congealed, and been converted into stones, minerals, metals, clays, earth, &c. and so in process of time lost their fluidity, and that the earth itself waxes old almost in the same manner as animals and vegetables do; its moisture gradually decaying or wasting, either into air, and from thence into æther; or else by degrees the parts communicating their motion to the fluid æther, grow immoveable and hard. Therefore if it be probable that the parts of the earth have been formerly softer and more yielding, how much more powerful might earthquakes then be in breaking, raising, overturning, and otherwise changing the superficial parts of the earth: besides they might be more frequent before the fuels of the subterraneous fires were much spent; for that those do also waste and decay, is evident from the extinction and ceasing of several vulcano's that have heretofore raged; which considerations may afford us sufficient arguments to believe that earthquakes have heretofore, not only been much more frequent and universal, but likewise much more powerful.

Corollaries deduced from the preceding Propositions.

I. THAT there may have been in past ages, whole countries, either swallowed up into the earth, or sunk so low as to be drowned by the coming in of the sea, or divers other ways quite destroyed; as *Plato*'s *Atlantis*, &c.

II. That there may have been as many countries new made and produced, by being raised from under the water, or from the hidden parts of the body of the earth, as *England*.

III. That there may have been divers species of things wholly destroyed and annihilated, and divers others changed and varied: for since we find that there are some kinds of animals and vegetables peculiar to certain places, and not to be found elsewhere; if such places have been swallowed up, 'tis not improbable but that those animal beings may have been destroyed with them; and this may be true both of aerial and aquatic animals: for those animated bodies, whether vegetables or animals,

mals, which were naturally nouriſhed or refreſhed by the air, would be deſtroyed by the water. And this I imagine to be the reaſon why we oft find the ſhells of divers fiſhes petrified in ſtone, of which we have now none of the ſame kind; as divers of thoſe ſnake or ſnail ſtones whereof great varieties are found about *England*, and dug out of the midſt of the very quarry, ſometimes, in *Portland*, of a prodigious bigneſs.

IV. That there may have been divers new varieties generated of the ſame ſpecies, and that by the change of the ſoil by which it was produced: for ſince we find that the alteration of the clime, ſoil and nouriſhment often produces a very great alteration in vegetables; 'tis not to be doubted but that alterations alſo of this nature may cauſe a very great change in the ſhape, and other accidents of an animated body. And this I imagine to be the reaſon we find divers kinds of petrified ſhells, of which kind we have none now naturally produced.

V. 'Tis not impoſſible but that there may have been a preceding learned age, wherein poſſibly as many things may have been known as are now, and perhaps many more, all cultivated and reduced to their higheſt pitch; and all theſe annihilated, deſtroyed, and loſt by ſucceeding devaſtations.

VI. 'Tis not impoſſible but that this may have been the cauſe of a total deluge, which may have brought on a deſtruction of all things then living in the air: for if earthquakes can raiſe the ſurface of the earth in one place, and ſink it in another, ſo as to make it uneven and rugged with hills and pits, it may, on the contrary, level thoſe mountains

tains again, and fill thoſe pits and reduce the body of the earth to its primitive roundneſs, and then the waters muſt neceſſarily cover all the face of the earth again, as it did at the beginning of the world, and by this means not only a learned age may be wholly annihilated, and no relicks of it left, but alſo a great number of the ſpecies of animals and plants.

VII. 'Tis not impoſſible but that ſome of theſe great alterations may have altered alſo the polar directions of the earth; ſo that what is now under the pole, or æquator, or any other degree of latitude, may have formerly been under another: for ſince 'tis probable that divers of theſe parts that have ſuch a quality, may have been tranſpoſed, 'tis not unlikely but that the æquatoreal axis of the whole may be alter'd by it, after the ſame manner, as we may find by experiments on a loadſtone, that the breaking off and tranſpoſing the parts of it, do cauſe a variation of the magnetic axis.

Of Earthquakes in the Leeward Iſlands.

THE greateſt objection againſt my theory of the varieties obſervable in the preſent ſuperficies of the earth, as cauſed by the power of earthquakes, or eruptions of fiery conflagrations inkindled in the ſubterraneous regions, is, I find, the want of hiſtory to confirm it. For that all places, countries, ſeas, rivers, iſlands, *&c.* have all continued the ſame for ſo long a time as we can reach backwards with any hiſtory. All *Greece*, and

and the *Grecian Iſlands*, *Italy*, *Ægypt*, *&c.* are all the ſame as they were 2000 years ſince, and therefore they were ſo from the creation, and will be ſo to the general conflagration; and as to the effects of earthquakes, firſt, they have happened but ſeldom; and, ſecondly, they have not produced any notable change, ſuch as I have ſuppoſed them to be the authors and efficients of. So that it ſeems but a bare conjecture, and without ground and foundation ſufficient to found and raiſe ſuch a ſuperſtructure of concluſions, as I have thereupon raiſed.

In anſwer to which, I ſhall not repeat here what I have formerly produced; but ſhall take notice of ſome particular inſtances which have happened within our own memory, and more particularly of the late inſtance which hath happened in the *Antilles*, of which we have an account in the *Gazette*, namely in that of *June* 30th and another in that of *June* 16th preceding, both which relations, tho' they are but ſhort and imperfect, as to what I could have wiſhed for and ſhall endeavour to obtain; yet, as they are, they will be found to contain many particulars which very much illuſtrate and confirm my conjectures. And tho' the particular effects were not ſo great as to equalize thoſe which I have ſuppoſed to have been the productions of former eruptions; ſuch as the raiſing of the *Alpes*, *Pyreneans*, *Apennine*, *Andes*, and the like mountains; or the making of new lands, iſlands, *&c.* or the ſinking of countries and drowning of iſlands, as the *Platonic Atlantis* and contiguous iſlands, yet if they be conſidered, they will be found to be of the

the ſame nature, and to differ only in magnitude, but not in eſſence.

The firſt account is dated from *Nevis*, *April* the 30th, (1690) in theſe words. "On *Sunday* the
" 6th inſtant, about five o'clock in the evening,
" was, for ſome minutes, heard a ſtrange hollow
" noiſe, which was thought to proceed from the
" great mountain in the middle of the iſland, to
" the admiration of all people; but immediately
" after, to their great amazement, began a mighty
" earthquake; with that violence, that almoſt all
" the houſes in *Charles Town*, that were built of
" brick or ſtone, were, in an inſtant, levelled with
" the ground, and thoſe built with timber ſhook,
" that every body made what haſte they could to
" get out of them. In the ſtreets the ground in
" ſeveral places clove about two foot aſunder, and
" hot ſtinking water ſpouted out of the earth a
" great height. The ſea left its uſual bounds
" more than a third of a mile, ſo that very large
" fiſh lay bare upon the ſhore, but the water pre-
" ſently returned again: and afterwards the ſame
" ſtrange motion happened ſeveral times, but the
" water retired not ſo far as at firſt. The earth
" in many places was thrown up in great quanti-
" ties, and thouſands of large trees went with it,
" which were buried and no more ſeen. 'Tis
" uſual at almoſt every houſe to have a large ciſ-
" tern, to contain the rain water, of about nine
" or ten foot deep, and fifteen or twenty foot di-
" ameter; ſeveral of which, with the violence of
" the earthquake, threw out the water eight or
" ten foot high; and the motion of the earth all

" over

" over the iſland was ſuch, that nothing could be
" more terrible. In the iſland of St. *Chriſtopher*
" (as ſome *French* gentlemen who are come hither
" to treat about the exchange of priſoners do re-
" port) there has likewiſe been an earthquake,
" the earth opening in many places nine foot, and
" burying ſolid timber, ſugar mills, *&c.* and
" throwing down the Jeſuits college, and all other
" ſtone buildings. It was alſo in a manner as
" violent at *Antego* and *Montſerrat*; and they had
" ſome feeling of it at *Barbadoes.* Several ſmall
" earthquakes have happened ſince, three or four
" in 24 hours; ſome of which made the biggeſt
" rocks have a great motion, but we are now in
" great hopes there will be no more."

This is the whole of the relation from *Nevis*: but the other account from *Barbadoes*, of the 23d of *April*, takes notice of other particulars than what are mentioned in this letter: the printed account is as follows. " About three weeks ſince there
" were felt moſt violent earthquakes in the *Leeward*
" *Iſlands* of *Montſerrat*, *Nevis* and *Antego*; in the
" two firſt no conſiderable miſchief was done,
" moſt of their buildings being of timber; but
" where there were ſtone buildings, they were ge-
" nerally thrown down, which fell very hard in
" *Antego*, moſt of their houſes, ſugar mills, and
" wind mills being of ſtone. This earthquake
" was felt in ſome places of this iſland, but did
" no manner of hurt to men or cattle; nor was a-
" ny loſt in the *Leeward Iſlands*, it happening in the
" day.time. It is reported to have been yet more
" violent in *Martinico*, and other *French* iſlands,

 " and

" and ſeveral ſloops which came from *Nevis* and
" *Antego*, paſſing between St. *Lucia* and *Martinico*,
" felt it at ſea; the agitation of the water being
" ſo violent, that they thought themſelves on rocks
" and ſhelves, the veſſels ſhaking as if they would
" break in pieces. And others paſſing by a rock
" and uninhabited iſland, called *Rodunda*, found
" the earthquake ſo violent there, that a great
" part of that rocky iſland ſplit and tumbled into
" the ſea, and was there ſunk, making a noiſe as
" of many cannon, and a very great cloud of duſt
" aſcending into the air at the fall. Two very
" great comets have lately appeared in theſe parts
" of the world, and in an hour and a quarter's
" time the ſea ebbed and flowed to an unuſual de-
" gree, three times."

In theſe relations are many conſiderable effects produced which will much confirm my former doctrine about earthquakes. And firſt, it is very remarkable, that this earthquake was not confined to a ſmall ſpot or place of the earth, ſuch as the eruption of *Ætna* or *Veſuvius* out of one mouth, but it extended above five degrees, or 350 miles in length, from *Barbadoes* to St. *Chriſtopher*'s, and poſſibly much farther: and tho' there might not be opportunities of noticing the effects in all places of the ſea where it might have been felt; yet by the few inſtances related, we may gueſs that its effects might be very conſiderable, and ſenſible a great way in breadth under the ſea; for we find that the ſuccuſſions were felt by veſſels ſailing over ſome parts of the ſea ſo affected, and thoſe ſo violent, as if the veſſels had ſtruck upon rocks; which could

could be from nothing elſe but the ſudden riſing of the bottom of the ſea, which raiſed the ſea alſo with it, like water in a tub or diſh: and that this was of that nature, does further appear by the unuſual tides at *Barbadoes* mentioned in the laſt relation, which in all probability was nothing elſe but waves propagated from the places where the ground underneath, and the ſea above, had been, by the concuſſions of the earthquake, raiſed upwards. This appears alſo farther by the receſs of the ſea at *Nevis*; for the whole iſland being raiſed by the ſwelling or eruption of the vapour or fire underneath, made the ſea run off from the ſhores, 'till it ſettled down again, after the vapour had broken its way out thro' the clefts that were made by thoſe ſwellings. From all which particulars, and ſeveral others, 'tis manifeſt, that the ſpace of earth raiſed or ſtruck upwards by the impetuoſity of the ſubterraneous powers, was of great extent, and might far exceed the length of the *Alpes* or the *Pyreneans*, &c.

Another notable particular is the receſs of the ſea from the ſhore, and the leaving the fiſh upon the ſo raiſed bottom: and tho' this part ſoon after ſunk again, ſo that the ſea returned to its former bounds; yet if ſome other parts of the ſubterraneous ground had filled up the new made cavity, or had ſo tumbled as to ſupport the ſo raiſed parts, then it would have left ſome ſuch kind of tract as is now in *Virginia*, where, for many miles in length, the lowland is nothing but ſea ſand and ſhells, which have been, in all probability, ſo raiſed into the air, and there ſupported and kept from ſinking down

down again into the ſea. There can be no doubt that the ſhells taken up from this tract did belong to fiſh of their kind, they remaining perfect ſcallop ſhells to this day.

A third remarkable particular, is the burying and covering of thouſands of trees by the earth which was thrown up by the eruption. This is a plain inſtance how trees found buried in many parts of *England* may have come to be ſo depoſited, probably at a time before any writings or records were kept here; or, if ſince the *Roman* conqueſt, the neighbouring inhabitants might have periſhed in the cataſtrophe, whilſt thoſe at a diſtance might not think themſelves ſufficiently intereſted in tranſmitting the account to poſterity. *Ariſtotle* ſpeaking of the like events [q], ſays, "Now, becauſe "many of theſe changes happen but ſlowly, in "compariſon to the quickneſs and ſhortneſs of the "life of man, therefore they are hardly taken notice "of, a whole generation having paſſed away be-"fore ſuch changes have come to perfection. O-"ther cataſtrophies that have been more quick, "have been forgotten, by reaſon that ſuch as eſ-"caped them were removed to ſome other parts, "and there the memory of them was ſoon loſt; at "leaſt a longer tract of time did quite obliterate "the remembrance of them, and the tranſplanting "and tranſmigration of people from place to place "much contributed thereto." This is made plain by the little remembrance there was found in *America* of their preceding eſtate, when they were firſt viſited by the *Spaniards* and other *Europeans*.

[q] Meteor. lib. i. cap. 14.

A

A fourth particular remarkable in theſe relations, is the chopping and cleaving of the earth and rocks, and the ſpouting of ſtinking water out of them to a great height, as alſo of ſmoke or duſt; which ſerves to explain the reaſon and cauſes of the flaws and veins in marbles and other ſtones: for by the power and violence of the ſubterraneous heavings or ſuccuſſions, the ſtony quarries become broken, flawed and cleft, and ſubterraneous mineral waters impregnated with ſaline, metalline, ſulphureous, or other ſubſtances are driven into them and fill them up, which having petrifying qualities, do, in proceſs of time, petrify in thoſe clefts, and thereby form a ſort of ſtony veins, of different colour, hardneſs, and other qualifications, than what the parts of the broken quarry had before, and oft-times incloſe divers other ſubſtances, by their petrifying quality, which have happened to fall into thoſe clefts; and thence ſometimes there are found ſhells petrified in the middle of a vein, and other ſubſtances. Theſe clefts or chaps happen not only upon the land, but even under the ſea; ſo that not only the ſea water may deſcend and fill them up, but may carry with it ſand, ſhells, mud, and divers other matters from the bottom of the ſea, that then lay above it; there to be in proceſs of time changed into ſtone, ſomewhat of the nature of that which has been ſo cleft.

Fifthly, 'tis worth noting, that this earthquake happened at ſo great a diſtance from the main land and great continent, and that the noiſe of the ſame was firſt obſerved to begin at the great mountain in the middle of the iſland of *Nevis*, not but that

in other parts it might have begun ſooner or at o-
ther times; from which I infer: firſt, that it ſeems
probable that this great mountain may have been
firſt produced by ſome ſuch power, and ſo have
great cavities within its bowels formed by ſuch a
preceding eruption, the diſlocated parts not returning each to his own place. And next, that it may
hence ſeem probable, that ſome ſuch preceding
earthquake, perhaps more violent for the firſt
time, might not only be the cauſe of raiſing this
mountain, but of lifting up from the bottom the
whole iſle, nay poſſibly of all the iſlands of the
Antilles, ſince one ſeems as poſſible as the other,
and the northermoſt of them all ſeems to hint as
much, if conſidered in the map: beſides, there
ſeems to be many inſtances of a like nature, as in
the *Canaries*, *Teneriffe* is a remarkable character of
ſuch a ſuppoſition; to which may be added *Del
Fuego* and *Madeira*; *Sicily*, *Strombulo* and *Lipary* in
the *Mediterranean*; *Iceland* in the *North Sea*; *Maſ-
carenos* near *Madagaſcar*; with the many iſlands of
the *Archipelago*, which though they have now no
great ſign of burning mountains, yet to this day
earthquakes are very frequent there, and ancient
traditions do preſerve ſomewhat of the memory of
very great alterations that have happened from
ſuch cauſes. And I do not queſtion but that all
iſlands which lie far in the ſea, would plainly manifeſt, if they were thoroughly examined, whence
they have proceeded, and this by characters of nature's writing, which to me are far beyond any o-
ther record whatſoever.

Sixthly,

Sixthly, 'tis very remarkable that the Iſle of *Rodunda*, being all an uninhabited rock, was ſplit, and part of it tumbled down and ſunk into the ſea, with a noiſe as of many cannon; ſending up at the ſame time a great cloud of duſt, as they term it, which in all probability was alſo mingled with ſmoak: which puts me in mind of the phænomena I obſerved lately, when the powder mill and magazine at *Hackney* blew up; for beſides the very great noiſe of the blow I heard, being within a mile of it in the fields, I obſerved immediately a great white cloud of ſmoak to riſe in a body to a great height in the air, and to be carried by the wind for two miles and better, without diſperſing or falling down, and perfectly reſembling the white ſummer clouds. From theſe phænomena of the earthquake it ſeems very probable, that it proceeded from ſuch ſubterraneous inkindling as reſembles gun-powder, both by the noiſe it yielded, and its ſuddenneſs of firing, and its powerful expanſion when fired. Next, the ſplitting of the rocky iſland proves its power to be very great, which is proved yet farther by the blow and ſtrokes it communicated to the ſea, and to the ſhips that ſailed on it; for no ſlow motion whatever could have communicated ſuch a concuſſion through the water to the veſſels upon it, but it muſt be as ſudden as that of powder; for if it had been a gradual riſing from the bottom, the ſea would gradually have ran off from it, and upon its ſinking again have gradually returned, and the veſſels on it would only have been ſenſible, at moſt, but of a current or running of the water, to or from the place of ſinking

ing or rising, somewhat like the effect which happened at *Nevis*; which plainly shews, that besides the sudden strokes or concussions, there was also a considerable rising and sinking of the whole island: but what I principally note under this head, is, a good part of the island's tumbling and sinking into the sea, which shews how many parts of the earth come to be buried, and displaced from their former situations, and thence how ships anchors, bones, teeth, *&c.* that have been digged up from great depths, may have been there buried.

Seventhly, 'tis remarkable also, that this eruption sent up into the air vast clouds of dust and smoke, which for the most part must soon fall down again into the sea, or contiguous parts of the island. This will give a probable account how the layers of the superficial parts of the earth may come to be made; for the most part of this dust must come down to the bottom first, and settle to a certain thickness, and make a bed of gravel, and then will follow beds of coarse sand, then beds of a finer sand, and last, of clays or moulds of several sorts. Again, much of that which fell upon the higher parts of the island, will, by the rivers, be washed down into the vales, and there produce the like beds or layers of several kinds, and so bury many of the parts that were before on the surface. Thus plants and vegetable substances may come to be buried, and the bones and teeth of the carcasses of dead animals: these may also sometimes be buried under beds or crusts of stone, when the parts that thus make the layers, chance to be mixed with such subterraneous substances as carry with them

them a petrifying quality. I could heartily wiſh that ſome care were taken, that a more particular account were procured of theſe earthquakes whilſt their effects were freſh in memory, that they might be recorded and added to the collections of natural hiſtory: and for the ſame end it were deſireable to kuow what former earthquakes have been taken notice of in theſe iſlands, as *Jamaica*, *Cuba*, *Hiſpaniola*, *Porto Rico*, &c. for the circumſtances of ſuch accidents, if they be not collected and recorded whilſt the ſpectators are in being, are ſoon forgotten, and loſt, or not regarded by ſucceeding generations, as *Ariſtotle* has well obſerved in a chapter I before quoted.

Why Iſlands and Sea Coaſts are moſt ſubject to Earthquakes.

WHAT is moſt remarkable in theſe earthquakes in the *Leeward Iſlands*, is, that they have all happened to places not far diſtant from the ſea, or even under the ſea itſelf, though the eruptions have been, for the moſt part, on the land. So that there ſeems to be ſome reaſon to conjecture, as Signior *Bottoni* does in his *Pyrologia Topographica*, that the ſaline quality of the ſea water may conduce to the production of the ſubterraneous fermentation with the ſulphureous minerals there placed, which an experiment lately exhibited before the Royal Society, makes ſtill more probable; for it appeared that the mixing of ſpirit of ſalt with iron, did produce ſuch a fermentation as raiſed a vapour or ſteam which by an actual flame was immediately fired like gun-powder, and if incloſed would, in all probability, have had a like effect of raiſing and diſperſing of thoſe parts that bounded and impriſoned it. Now, the melted matter vomited out of *Ætna* in the year 1669, was very much like to melted or caſt iron, and I doubt not but that there may be much of that mineral in it; beſides the foot of that mountain extends even to the very ſea, and in all probability may have caverns under the ſea itſelf, which is argued alſo from the ſimultaneous conflagration of *Strombolo* and *Lipary*, iſlands conſiderably diſtant from it by ſea, where it is generally believed that there may be cavernous paſſages between them, by which they communicate; ſo that ſometimes it begins in *Ætna*, and is communicated to *Strombolo*, and reciprocally communicated to *Mongibel*.

This may poſſibly afford a probable reaſon why iſlands are now more ſubject to earthquakes, than continents and inland parts; and indeed how ſo many iſlands came to be diſperſed up and down in the ſea; for that theſe fermentations may have been wrought up in ſubmarine parts of the earth, and being ripe may have taken fire, and ſo have had force enough to raiſe a ſufficient quantity of the earth above it, to make its way through the ſea, and there gain a vent, as that of the *Canaries* did in the year 1639, which, if ſufficiently copious, may produce an iſland, as that did for a time, but has ſince again ſunk under the ſurface of the ſea. But the iſland of *Aſcenſion*, which by all appearance
ſeems

ſeems to have been produced the ſame way, ſtill remains a witneſs to prove this hypotheſis. Like teſtimonies are the iſland and *Pike* of *Teneriffe*, *Hecla* of *Iceland*, *Bearenberg* of *John Mayens* or *Trinity Iſland*, *Del Fuego* of the *Cape Verd* iſlands, *Ternate* of the *Moluccas*, *Maſcarenas*, ſome about *Madagaſcar*, and the *Antilles* or *Caribbees*. And tho' the fires be extinct in many of the other iſlands, yet 'tis obſervable that the prodigious high mountains or ſugar-loaf pikes do yet remain as marks of what they had been heretofore; ſo the *Pike* of *Fayal* among the *Terceras*, and the whole iſland of St. *Helena* and ſeveral about *Madagaſcar* and in the *Eaſt Indies*, and the *Antilles*, and that of St. *Martha* mentioned by *Dampier*, ſeem plain evidences of the original cauſes of them all, tho' at various periods of time.

Of the Cauſes of EARTHQUAKES.

THE materials that ſerve to produce earthquakes, I conceive to be ſomewhat analogous to the materials of gun-powder; not that they muſt neceſſarily be the very ſame, either as to the parts, or as to the manner or order of compoſition, or as to the way of inkindling or accenſion; for that as much the ſame effect may be produced by differing agents, ſo the methods and order of proceeding may be altogether as various: a clear inſtance of which we have in the phænomena of lightning, wherein we may obſerve that the effects are very like to the effects of gun-powder. For we have firſt the flaſh of light, which is very ſudden, very bright, and of very ſhort continuance, being almoſt momentaneous. Next we may obſerve the violence of the crack, which is likewiſe momentaneous, if it be ſingle, but if there be many particular accenſions which contribute to this effect, and thoſe made at ſeveral diſtances, then the thunder is heard longer than the duration of the flaſhes, as I conceive, from two cauſes; firſt, for that thoſe flaſhes that are farther diſtant, have their thunder a longer time in paſſing to the ear, than thoſe which are nearer; becauſe that, though the motion be almoſt inſtantaneous, yet the motion requires a ſenſible time to paſs a ſenſible ſpace, and the times are proportionably longer, as the ſpaces paſſed are greater. But a ſecond cauſe of the duration of thunder, I imagine, proceeds from echoes that are rebounded, both from parts of the earth, and parts of the air, as from charged clouds; of both which I am ſenſibly aſſured, having obſerved the ſame effects produced by the echoing and rebounding of the ſound of a piece of ordnance. But thirdly, we have alſo the power and violence of the force of the fire and expanſion, in firing ſeveral combuſtibles, in ſuddenly melting of metals and other materials, otherwiſe difficult and ſlow enough to be made to flow; in rending, tearing throwing down, and deſtroying whatever ſtands in its way, *&c.* and yet after all, that which cauſes theſe, and many other ſtrange effects reſembling thoſe of gun-powder, ſeems to be nothing but a vapour or ſteam, mixed with the body of the air, which is kindled, not by any active fire, but by a kind of fermentation, or inward working of the ſaid

ſaid vapour. Again, we find that the *Pulvis Fulminans*, as 'tis called, which has ſome of its materials differing from that of common powder; as alſo *Aurum Fulminans*, which differs ſtill more, both as to its materials, and its way of kindling, have yet moſt of the ſame effects with gun-powder, both as to the flaſhing and thundering noiſe, and as to the force and violence: ſo that theſe are differing in many particulars, and yet produce much the ſame effects; whence 'tis probable, that what is the cauſe of earthquakes, and ſubterraneous thundring, lightning, and violent expanſion, as I may ſo call thoſe phænomena obſervable in thoſe criſes of nature, may be in divers particulars, different from every one of theſe, both as to the materials, and the form and manner of accenſion; and yet, as to the effects, they may be very analogous and ſimilar: ſo that 'tis but one operation in nature, and that which cauſes the effect in one cauſes the effect in all the reſt; the outward appearances of the different materials, and the differing way of operating, being nothing but their different modes of acting their ſeveral parts, which, when they have done, they are at an end, and there muſt be a new ſet of actors to do the ſame thing again. So the materials that make the ſubterraneous fire, flame, or expanſion, call it by which name you pleaſe, is conſumed and converted into another ſubſtance, unfit to produce any more the ſame effect; and if the conflagration be ſo great as to conſume all the preſent ſtore, you may ſafely conclude that place will no more be troubled with ſuch effects; but if there be left relicts, either already

ready fit and prepared, though ſheltered from accenſion, by ſome interpoſing incombuſtible materials, or that there be other parts not thoroughly ripe and ſufficiently prepared for ſuch accenſion, then a concurrence of after cauſes may repeat the ſame effects, and that *toties quoties*, 'till all the mine be exhauſted; which I look upon both poſſible and probable, nay neceſſary, becauſe I find it to be the general method of nature, always to be going forward in a progreſs of changing all things from the ſtate in which it finds them. All things, as they proceed to their perfection, ſo they proceed alſo to their diſſolution and corruption, as to their former ſtate; and where nature repeats the proceſs, 'tis always on a new individual.

Now tho' it may be objected of the material production of lightning, that notwithſtanding it ſeems to be all kindled and burnt off by the flaſh, yet after ſome time the ſame is again renewed, and ſo from time to time; and therefore as one operation deſtroys and conſumes it, ſo another generates and reproduces it, and thence it ſeems probable that the ſame may be done in the ſubterraneous regions, ſo that there would be little reaſon to ſuppoſe that former earthquakes ſhould have been greater than thoſe obſerved in the preſent age: I would anſwer, that tho' it ſeems plain that the matter of lightning is renewed, yet I conceive that to be only by new emanations from the proper minerals in the bowels of the earth, and not becauſe the ſame ſubſtance burnt off in the lightning, is again reſtored to its former ſtate, and fitted for a ſecond accenſion; for though a previous digeſtion of the

ſteams

ſteams may be neceſſary, yet that only prepares it, but it muſt be ſome proper mineral that muſt furniſh the materials. And the ſame is more evident in vulcano's, which are there only obſerved to break forth where there is plenty of brimſtone and other combuſtible ſubſtances; for were it only a continual new generation of materials for fire, then I ſee no reaſon why thoſe *incendiums* ſhould not be equally frequent and great in all places. It follows therefore, that it muſt be cauſed, not by the renovation of the fewel, but from the duration of the mines or minerals that ſupply fit materials, and conſequently, that when thoſe ſhall be quite conſumed, then, and not till then, will the fire go quite out. Nay, that there are ſome ſuch inſtances of preceding vulcano's, which have heretofore burned, and are now quite ſpent, may be concluded from the *Pike* of *Teneriffe*, which ſeems to carry the ſtrongeſt evidence of having been formerly a burning mountain; and the iſland of *Aſcenſion* ſeems to be another ſuch inſtance. All which conflagrations are the ſeveral ſymptoms of the progreſs of nature in her determined courſe and method.

I cannot therefore perceive any abſurdity in thinking or aſſerting, that this globe of the earth is in a ſtate of progreſſion from one degree of perfection to another, in as much as it is the progreſs of nature; and at the ſame time that it may be conceived in a ſtate of corruption and diſſolution, in as much as it is continually changed from its preceding ſtate to a new one, which may be, upon ſome accounts, conſidered as more perfect, tho' upon

upon others it may be reckoned corrupting, and tending to its final diſſolution; and as 'tis moſt certain that it is continually older in reſpect of time and duration, ſo I conceive alſo that it grows older, as to its conſtitution and powers; and that there have been many more effects produced by it in its more juvenile ſtate, than it can now produce in its more ſenile, particularly as to earthquakes and eruptions; for to me it ſeems beyond a doubt that there have been in preceding ages many of theſe which have infinitely ſurpaſs'd any of later years, or indeed all that we have any certain account of in hiſtory. A notice of ſome ancient traditions concerning a very great one, ſeems to be preſerved in the mythological hiſtory of *Phaeton*; of which *Plato* alſo tells us, that the *Ægyptians* had a more perfect account, than ever the *Greeks* were maſters of, who, at beſt, as to hiſtories of preceding ages, were by the prieſts of *Egypt* accounted boys and children. In which caſe we are to diſtinguiſh between hiſtories of matters of fact, and thoſe of opinion; and *Plato* hints as much in mentioning the relation. The matters of fact ſeem to have been the conflagration of many parts of the earth at once, and thoſe the moſt eminent, ſuch as the mountains, it being probable that this was the time of their production. We are not to conclude that ſuch huge mountains as the *Andes*, *Caucaſus*, *Atlas*, &c. could never be produced by means of earthquakes and eruptions, becauſe we do not now find inſtances of effects of the ſame grandeur, in this age, or in others of which we have ſome tolerable account; ſince in remoter times there has been much

much greater plenty of proper minerals, which were then consumed, and whose relicts are now but small, and probably not so apt for conflagration, nor so strong in their operations; besides many that were left, may have been since petrified, or converted into other substances, wholly unfit for the foment or fewel of such kinds of fires.

Petrification is a symptom of very old age, as plenty of spirituous, unctuous and combustible or inflammable juices and moisture is a sign of youth. Fluidity is an inseparable concomitant of what we call spirituous substances; and 'tis the plenty of those that makes both plants and animals to flourish in their youthful state, and the consumption and lack of them that make them decay and grow old, stiff, dry, rough, and shriveled; all which marks may plainly be discovered also in the body of the earth; and I am apt to think would be much more evident, if we could be truly informed of the younger condition thereof: I have very good reason to believe that times have been when it had a much smoother, softer and succous skin than now; when it abounded more with spirituous substances, when all its powers were strong and vegete, without any of its present scars, asperities and stiffness: and tho' some may possibly think all these conceptions groundless, and merely conjectural, yet I may in good time manifest, that there are other ways of coming at the discovery of many truths, than what have been hitherto made use of to this purpose, which yet are not less capable of proof and confirmation, than histories and records are from coins, inscriptions or monuments.

 To

To conclude. The aſſertion of the earth's growing old, cannot be looked upon either as a heterodoxical, or a ſchiſmatical one: the kingly prophet has an expreſſion which does plainly declare it, not only of the earth, but of the heavens too [r]. "Of old haſt thou laid the foundations of the "earth, and the heavens are the work of thy hands; "they ſhall periſh, but thou ſhalt endure, yea all "of them ſhall wax old like a garment, as a veſ-"ture ſhalt thou change them, and they ſhall be "changed." Which expreſſion is almoſt verbally repeated by the prophet *Iſaiah* [ſ]. "Lift up your "eyes to heaven, and look upon the earth be-"neath; for the heavens ſhall vaniſh away like "ſmoak, and the earth ſhall wax old like a gar-"ment." Nay this expreſſion of the pſalmiſt is again *verbatim* repeated by the apoſtle to the *Hebrews* [t]. "And thou Lord in the beginning haſt "laid the foundation of the earth, and the heavens "are the work of thine hands: they ſhall periſh, "but thou remaineſt; and they all ſhall wax old "as doth a garment; and as a veſture ſhalt thou "fold them up; and they ſhall be changed." By all which it is evident at leaſt, that *David*, *Iſaiah*, and St. *Paul*, were all of this belief. I could produce many expreſſions to the like purpoſe, both in ſacred and prophane hiſtories of chriſtian and heathen writers, but thoſe I have quoted I ſuppoſe may be ſufficient to anſwer ſuch objectors.

[r] Pſalm cii. v. 25, 26. [ſ] Chap. li. v. 6. [t] Chap. i. v. 10, 11, 12.

EARTH-

EARTHQUAKES

Caufed by fome accidental obftruction of a continual fubterranean Heat.

I Suppofe that the fubterranean heat or fire, which is continually elevating water out of the abyfs to furnifh the earth with rain, dew, fprings and rivers, when it is ftopped in any part of the earth, and fo diverted from its ordinary courfe by fome accidental glut, or obftruction in the pores or paffages thro' which it ufed to afcend to the furface, becomes by this means preternaturally affembled, in a greater quantity than ufual, into one place; and therefore caufes a great rarefaction and intumefcence of the water of the abyfs, putting it into very great commotions and diforders; and at the fame time making the like effort on the earth, which is expanded upon the face of the abyfs; and that this occafions that agitation and concuffion of it, which we call an earthquake.

That this effort is in fome earthquakes fo vehement, that it fplits and tears the earth, making cracks and chafms in it fome miles in length, which open at the inftant of the fhock, and clofe again in the intervals betwixt them; nay, 'tis fometimes fo extreamly violent, that it plainly forces

the

the ſuperincumbent ſtrata; breaks them all throughout, and thereby perfectly undermines and ruins the foundation of them; ſo that theſe failing, the whole tract, as ſoon as ever the ſhock is over, ſinks down to rights into the abyſs underneath, and is ſwallowed up by it, the water thereof immediately riſing up, and forming a lake in the place where the ſaid tract before was.

That ſeveral conſiderable tracts of land, and ſome with cities and towns ſtanding upon them; as alſo whole mountains, many of them very large, and of great height, have been thus totally ſwallow'd up.

That this effort being made in all directions indifferently; upwards, downwards, and on every ſide; the fire dilating and expanding on all hands, and endeavouring proportionably to the quantity and ſtrength of it, to get room, and make its way through all obſtacles, falls as foul upon the water of the abyſs beneath, as upon the earth above, forcing it forth which way ſoever it can find vent or paſſage, as well through its ordinary exits, wells, ſprings, and the outlets of rivers; as thro' the chaſms then newly open'd; through the *camini* or ſpiracles of *Ætna*, or other near vulcano's; and thoſe *hiatus* at the bottom of the ſea, whereby the abyſs below opens into it, and communicates with it.

That as the water reſident in the abyſs, is in all parts of it, ſtored with a conſiderable quantity of heat, and more eſpecially in thoſe where theſe extraordinary aggregations of this fire happen, ſo likewiſe is the water which is thus forced out of it;

it; inſomuch, that when thrown forth, and mixed with the waters of wells, of ſprings, of rivers, and the ſea, it renders them very ſenſibly hot.

That it is uſually expelled forth in vaſt quantities, and with great impetuoſity, inſomuch that it hath been ſeen to ſpout out of deep wells, and fly forth at the tops of them, upon the face of the ground; with like rapidity comes it out of the ſources of rivers, filling them ſo of a ſudden, as to make them run over their banks, and overflow their neighbouring territories, without ſo much as one drop of rain falling into them, or any other concurrent water to raiſe and augment them.

That it ſpews out of the chaſms, opened by the earthquake in great abundance; mounting up in mighty ſtreams to an incredible height in the air, and this oftentimes at many miles diſtance from any ſea.

That it likewiſe flies forth of the volcano's in vaſt floods, and with wonderful violence: that 'tis forced through the *hiatus's*, at the bottom of the ſea, with ſuch vehemence, that it puts the ſea immediately into the moſt horrible diſorder and perturbation imaginable, even when there is not the leaſt breath of wind ſtirring, but all till then calm and ſtill; making it rage and roar with a moſt hideous and amazing noiſe, raiſing its ſurface into prodigious waves, and toſſing and rowling them about in a very ſtrange and furious manner; overſetting ſhips in the harbours, and ſinking them to the bottom, with many other like outrages.

That 'tis refunded out of theſe hiatus's in ſuch quantity alſo, that it makes a vaſt addition to the water of the ſea; raiſing it many fathoms higher than ever it flows in the higheſt tides, ſo as to pour it forth far beyond its uſual bounds, and make it overwhelm the adjacent country; by this means ruining and deſtroying towns and cities, drowning both men and cattle; breaking the cables of ſhips, driving them from their anchors, bearing them along with the inundation ſeveral miles up into the country, and there running them aground; ſtranding whales likewiſe, and other great fiſhes, and leaving them, at its return, upon dry land.

That theſe phænomena are not new, or peculiar to the earthquakes which have happen'd in our times, but have been obſerved in all ages, and particularly theſe exorbitant commotions of the water of the globe.

This we may learn abundantly from the hiſtories of former times; and 'twas for this reaſon that many of the ancients concluded rightly enough, that they were cauſed by the impulſes and fluctuation of water in the bowels of the earth; and therefore they frequently called *Neptune*, Σεισίχθων, as alſo, Ἐνοσίχθων Ἐνοσίγαιος, and Τινακτορογαίης; by all which epithets they denoted his power of ſhaking the earth.

They ſuppoſed that he preſided over all water whatever, as well as that within the earth, as the ſea, and the reſt upon it; and that the earth was ſupported by water, its foundations being laid thereon; on which account it was that they beſtowed upon him that cognomen Γαιήοχος, or ſupporter

porter of the earth, and that of Θεμελιῦχος, or the sustainer of its foundations.

They likewise believed, that he having a full sway and command over the water, had power to still and compose it, as well as to move and disturb it, and the earth, by means of it; and therefore they also gave him the name of Ασφάλιος, or, the establisher; under which name several temples were consecrated to him, and sacrifices offered, whenever an earthquake happened, to pacify and appease him; requesting that he would allay the commotions of the water, secure the foundations of the earth, and put an end to the earthquake.

That the fire itself, which being thus assembled and pent up, is the cause of all these perturbations, makes its own way also forth, by what passages soever it can get vent; through the spiracles of the next volcano's, through the cracks and openings of the earth abovementioned, through the apertures of springs, especially those of the *thermæ*, or any other way that it can either find or make; and being thus discharged, the earthquake ceaseth, till the cause returns again, and a fresh collection of this fire commits the same outrages as before.

That there is sometimes in commotion, a portion of the abyss of that vast extent, as to shake the earth incumbent upon it, for so very large a part of the globe together, that the shock is felt the same minute precisely, in countries that are many hundreds of miles distant from each other; and this, even tho' they happen to be parted by the sea lying betwixt them; there wants not instances of such an universal concussion of the whole

 globe,

globe, as muſt needs imply an agitation of the whole abyſs.

That though the abyſs be liable to theſe commotions in all parts of it, and therefore no country can be wholly exempted from the effects of them; yet theſe effects are no where very remarkable, nor are there uſually any great damages done by earthquakes, except only in thoſe countries which are mountainous and conſequently ſtony, and cavernous underneath, and eſpecially where the diſpoſition of the *ſtrata* is ſuch, that thoſe caverns open into the abyſs, and ſo freely admit and entertain the fire, which aſſembling therein, is the cauſe of the ſhock; it naturally ſteering its courſe that way where it finds the readieſt reception, which is towards thoſe caverns, this being indeed much the cauſe of damps in mines. Beſides, that thoſe parts of the earth which abound with *ſtrata* of ſtone, or marble, making the ſtrongeſt oppoſition to this effort, are the moſt furiouſly ſhattered, and ſuffer much more by it than thoſe which conſiſt of gravel, ſand, and the like laxer matter, which more eaſily give way, and make not ſo great reſiſtance; an event obſervable not only in this, but all other exploſions whatever.

But above all, thoſe countries which yield great ſtore of ſulphur and nitre, are by far the moſt injured and incommoded by earthquakes; thoſe minerals conſtituting in the earth, a kind of natural gun-powder, which taking fire upon this aſſembly, and approach of it, occaſions that murmuring noiſe, that ſubterranean thunder, which is heard rumbling in the bowels of the earth during

earth-

earthquakes, and by the aſſiſtance of its exploſive power, renders the ſhock much greater, ſo as ſometimes to make miſerable havock and deſtruction.

And 'tis for this reaſon, that *Italy*, *Sicily*, *Anatolia*, and ſome parts of *Greece*, have been ſo long, and ſo often alarm'd and haraſs'd by earthquakes; theſe countries being all mountainous and cavernous, abounding with ſtone and marble, and affording ſulphur and nitre in great plenty.

That *Ætna*, *Veſuvius*, *Hecla*, and the other volcano's, are only ſo many ſpiracles, ſerving for the diſcharge of this ſubterranean fire, when 'tis thus preternaturally aſſembled. That where there happens to be ſuch a ſtructure, and conformation of the interior parts of the earth, as that the fire may paſs freely and without impediment, from the caverns wherein it aſſembles unto thoſe ſpiracles, it then readily and eaſily gets out, from time to time, without ſhaking or diſturbing the earth; but where ſuch communication is wanting, or paſſages not ſufficiently large and open, ſo that it cannot come at the ſaid ſpiracles without firſt forcing and removing all obſtacles, it heaves up, and ſhocks the earth, with greater or leſſer impetuoſity, according as the quantity of fire thus aſſembled is greater or leſs, till it hath made its way to the mouth of the volcano; where it ruſheth forth ſometimes in mighty flames, with great velocity, and a terrible bellowing noiſe.

That therefore, there are ſcarcely any countries that are much annoy'd with earthquakes, that have not one of theſe fiery vents, and theſe are conſtantly all in flames when any earthquake happens, they diſgorging that fire, which whilſt underneath, was the cauſe of the diſaſter; and were it not for theſe diverticula, whereby it gains an exit, 'twould rage in the bowels of the earth much more furiouſly, and make greater havock than now it doth.

So that tho' thoſe countries, where there are ſuch volcano's, are uſually more or leſs troubled with earthquakes; yet were theſe volcano's wanting, they would be more troubled with them, than now they are; yea, in all probability, to that degree, as to render the earth for a vaſt ſpace around them, perfectly uninhabitable.

In one word, ſo beneficial are theſe to the territories where they are, that there do not want inſtances of ſome which have been reſcued and wholly delivered from earthquakes by the breaking forth of a new volcano there; this continually diſcharging that matter, which being till then barricado'd up, and impriſoned in the bowels of the earth, was the occaſion of very great and frequent calamities.

That moſt of thoſe ſpiracles perpetually, and at all ſeaſons ſend forth fire, more or leſs; and tho' it be ſometimes ſo little, that the eye cannot diſcern it; yet, even then, by a nearer approach of the body, may be diſcovered a copious and very ſenſible heat continually iſſuing out.

A PHY-

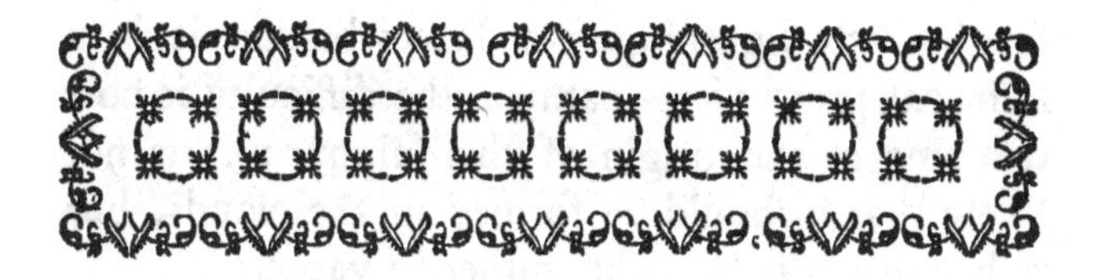

A

PHYSICO-CHYMICAL

EXPLANATION

OF

Subterraneous Fires, Earthquakes, &c.

MY intention is to give, by the means of a chymical operation, a ſenſible idea of what is tranſacted in the clouds when they are burſt open during a tempeſt, ſo as to produce lightning and thunder: but before I come to the experiment, it will be proper to ſay ſomething of the matter which is immediately concern'd in cauſing ſuch violent effects, and to examine into its nature and origin.

It cannot reaſonably be doubted that the matter of lightning and thunder is a ſulphur inflamed and diſcharged with prodigious rapidity. The ſulphury ſmell which lightning ever leaves behind it is a ſufficient proof of its nature: the difficulty is how to come at the origin of this ſulphur: it is not likely that it ſhould be formed in the clouds, but rather that it is brought thither in vapour.

To me it appears that the origin of the matter which produces thunder, is the ſame as that which cauſes earthquakes, hurricanes and ſubterraneous fires, *&c.* I have explained the cauſe of theſe grand commotions in my book of chymiſtry, on the occaſion of a particular preparation of iron called *Saffron of Mars*, which I publiſhed ſeveral years ago; and having ſince made ſeveral other experiments which ſerve to confirm what I have there advanced, I am willing to give a ſuccinct account of them all, the firſt of which is this.

I take a mixture of equal parts of filings of iron and ſulphur powdered; this I form into a paſte with water, and leave it to digeſt two or three hours, without fire, in which time it ferments and ſwells with a conſiderable heat; the fermentation cracks the paſte in divers places, and through the crevices there iſſue vapours, which indeed are but barely warm if the maſs be ſmall, but when it is conſiderable, as thirty or forty pounds, an actual flame comes forth.

The fermentation accompanied with heat, and even fire, which happens in this operation, proceeds from the penetration and violent friction which the acid points of the ſulphur exert upon the particles of the iron.

This ſingle experiment ſeems, to me, fully ſufficient for explaining after what manner fermentations,

tions, ſhocks and conflagrations are excited in the bowels of the earth, as happens in *Veſuvius*, *Ætna* and divers other places: for if iron and ſulphur happen to meet together, and are intimately united and penetrate each other, a violent fermentation muſt enſue, which will produce fire, as in our operation. But it is eaſy to prove, that in the mountains I have juſt now mentioned, there is both ſulphur and iron; for after the flames are over abundance of ſulphur is found on the ſurface of the earth; and in the paſſages through which the fire has paſſed, are diſcovered ſubſtances like thoſe which are ſeparated in our forges.

The following are experiments which I have made ſince the laſt edition of my book, and which confirm the former and ſtrengthen my argument.

I put of the ſame mixture of iron and ſulphur in different quantities into tall narrow pots, where I could compreſs the matter cloſer than before. Strong fermentations and ignitions enſued, and the matter was rais'd with a degree of violence, and part of it ſcattered round the pots.

In the ſummer ſeaſon I put fifty pounds of the ſame mixture into a large pot, which I cauſed to be placed in a hole dug in the earth in a field; it was covered with linen cloth, and with earth over that, about a foot thick. Eight or nine hours afterwards, the earth ſwelled, grew hot and cracked; then hot ſulphury vapours iſſued forth, and at length flames which widened the crevices, and ſcatter'd a black and yellow powder about the place: the earth continued hot a good while, which I removed after it was grown cold, and found nothing

thing in the pot but a weighty black powder, being the iron filings divested of part of the ſulphur: more earth might have been laid over the pot, but that it was ſuſpected that the matter would not kindle for want of air. This operation ſucceeds better in ſummer than in winter, on account of the heat of the ſun which excites a briſker motion in the particles of the iron and ſulphur.

It is then unneceſſary to look out any where elſe for the principle that puts ſulphurs in motion in mines, and ſets them on fire; their union with iron will produce perfectly this effect, in like manner as it produces it in our operations.

But here offers a difficulty; namely, that theſe vaſt ſubterraneous fermentations and conflagrations cannot have been produced without air: yet it can ſcarcely be apprehended how air ſhould find a paſſage to ſuch depths under ground.

To this objection I anſwer, that there are in the earth great numbers of chinks and paſſages which are not obvious to our ſight, eſpecially in hot countries, where ſuch ſubterraneous commotions moſt uſually happen: for the great force of the ſun-beams heating and calcining, as it were, the earth in divers places, forms crevices in it deep enough for the air to introduce itſelf.

Earthquakes ſeem to be occaſioned by a vapour, which having been generated in the violent fermentation of iron and ſuphur, is converted into a ſulphureous blaſt which forces a paſſage, and ruſhes wherever it can, raiſing and ſhaking the earth under which it moves. If this ſulphureous blaſt be continually kept confined ſo as not to be able to extricate itſelf through any aperture, the earthquake laſts a conſiderable time, and with ſtrong plunges, 'till its motion is become languid: but if it procures any paſſage to eſcape at, it ruſhes out impetuouſly, and creates what is called a hurricane, toſſing up the earth, forming abyſſes, tearing up trees by the roots, overſetting houſes; nor can men ſecure themſelves from its fury but by falling flat on their faces and cloſing their mouths, to ſave themſelves from being carried away, and to avoid breathing the hot ſuffocating ſulphury blaſt.

Subterraneous fires are owing to the ſame exhalation; the different effects which it produces ariſing from ſeveral cauſes; either from the greater abundance of the matter, and conſequently the ſtronger fermentation; or from a greater inlet of air; or from a number of chinks and crevices favouring the eſcape of the flames, carrying up clouds of aſhes along with them ſometimes ſufficient to cover whole villages, and ſuffocate or blind the inhabitants.

Ignes fatui, and the lights which appear on waters in hot countries, ſeem to derive their origin from the ſame cauſe; but the ſulphureous vapour having been but weak, and its motion impeded in filtering through ſand or water, it manifeſts itſelf only in a light lambent, ſpirituous and erratic flame, not having ſufficient matter to ſupport it long.

It is very probable that hot mineral waters, as thoſe of *Bourbon*, *Vichi*, *Balarue*, *Aix*, &c. do acquire their warmth from ſubterraneous fires, or hot ſulphury beds over which they glide. For when

when thoſe waters are left to ſettle, particles of ſulphur precipitate from them, and adhere to the ſides and bottoms of their containing veſſels.

Thoſe columns of water which are ſeen ſometimes at ſea, and threaten ſudden deſtruction to mariners, ſeem to be owing to theſe ſulphureous winds, driven rapidly up from under the ſea, after the like fermentations I have been treating of.

Theſe ſulphureous winds which occaſion hurricanes, are forced up with ſo great violence from under ground, that part of them are driven up even into the clouds, which conſtitutes the materials and cauſe of thunder: for this wind which contains an exalted ſulphur, is entangled among the clouds, and being there beaten backwards and forwards, and ſtrongly compreſſed, acquires motion ſufficient to ignite it, and produce lightning by burſting the cloud and darting itſelf forthwith with inconceivable rapidity: and this furious motion it is which produces the noiſe, which we hear, of thunder: for this ſulphureous blaſt iſſuing violently out from a ſtrait confinement, rudely attacks the contiguous air, and rowls through it with an extraordinary velocity, juſt as gun-powder out of the cannon wherein it was fired. It may be here ſaid, that a ſubtile nitre wherewith the air is at all times impregnated, is connected with the ſulphur of the thunder, and encreaſes the force of its motion and action; in like manner as when ſalt petre has been mixed with common brimſtone, it produces a far more violent effect in the rarefaction, than it is capable of by itſelf.

This

This ſulphureous wind of thunder, after rowling ſome time in the air, ſlackens its motion; on which account thunder is far more violent and dangerous the moment it is diſcharged from the cloud, than after it has performed ſome of its whirlings in the air, being in a very ſhort ſpace reduced to nothing, and leaving only a ſulphury ſtench behind it in the places through which it has paſs'd.

As to the thunder ſtones which the vulgar believes always to accompany lightning, their exiſtence may in my opinion well be queſtioned, and I verily believe there never was an inſtance of any ſuch thing: it is not however abſolutely impoſſible, that by a rapid aſcent of an hurricane to the clouds there may ſometimes be carried up with it ſome ſtony or mineral ſubſtances, which being ſoftened and melted together by heat, may form what is called a thunder ſtone: but ſuch ſtones are not found in places where it thunders; and if any ſuch ſhould be found, it would be more reaſonable to believe that it aroſe from a mineral ſubſtance melted and formed by the inflamed ſulphur of thunder in the earth itſelf, than to imagine that it was formed in the air or the clouds, and projected downwards with the thunder.

A difficulty ſtill remains; which is to know how the ſulphureous wind, which I have ſuppoſed to be the matter of thunder, comes to be kindled among the clouds, which conſiſt of water, and to be there compreſs'd without being extinguiſhed; for it ſhould ſeem that the water of the clouds ſhould

ſhould prevent the accenſion of the ſulphur; or at leaſt that it ſhould abſorb it when kindled.

To anſwer this difficulty, I ſay that ſulphur, being a pinguous ſubſtance, is not ſo liable to the impreſſion of water, as other matters are, and that it may be inflamed and burnt in water, like camphire and divers other exaltedly ſulphureous bodies. It muſt needs be, I own, that ſome part of this ſulphur being plunged into the maſs of water which conſtitutes clouds, will be extinguiſhed with a great detonation, like what happens when ſome ſolid red hot matter, as iron, is caſt into water: this detonation may poſſibly contribute to the noiſe of the thunder, but the other more ſubtile part, and the moſt diſpos'd to motion will be expell'd in a perfect ſtate of ignition. The following experiment will be a proof of my reaſoning.

Into a moderate ſized matras whoſe neck had been partly cut off, I put three ounces of good ſpirit of vitriol, and twelve ounces of common water; having warmed the mixture a little, I threw into it, at ſeveral times, an ounce or an ounce and half of iron filings, which produced an ebullition and white vapours; I preſented a lighted wax candle to the mouth of the matras, and the vapour inſtantly took fire with a very loud and violent fulmination; I repeated the application of the candle ſeveral times, and fulminations ſucceeded like the firſt, during which the matras was often filled with a flame which penetrated and circulated to the very bottom of the liquor, and ſometimes the flame laſted a conſiderable time in the neck of the veſſel.

There

There are ſeveral remarkable circumſtances in this operation. The firſt ebullition which happens on the throwing in the iron filings, proceeds from the ſolution of a portion of the iron by the ſpirit of vitriol; but to render the fumes and the ſolution the ſtronger, 'tis neceſſary to mix water with the ſpirit of vitriol, in the proportion mentioned; for if the ſpirit were pure and not diluted, and expanded with water, its points indeed would attack the iron, but they would be ſo embarraſs'd and compreſs'd together, that they would not have a freedom of motion ſufficient to produce any fulmination.

The ſecond is, that the liquor muſt be warmed a little to excite the points of the diſſolvents to penetrate the iron and raiſe fumes; but it muſt not be made too hot, for then the fumes would eſcape too faſt, and would only flame in the neck of the matras upon applying the candle, without any fulmination; for that noiſe ariſes from the ſulphureous part of the matter being kindled quite to the bottom of the matras, and meeting with an obſtacle to its riſing from the body of the water which it endeavours to eſcape through.

The third is, that the ſulphur which elevates itſelf in vapour and takes fire, muſt neceſſarily ariſe from the filings of iron alone, ſince neither the water, nor the ſpirit of vitriol, eſpecially the ſtronger ſort which I make uſe of, hold nothing of a ſulphureous or inflammable nature, as every one knows: it follows then that the ſulphur of the iron filings, having been rarefied and detached

by

by the ſpirit of vitriol, is exaled in a vapour extremely ſuſceptible of ignition.

The fourth, that the acid ſpirits of ſalt, ſulphur and alum produce in this operation, the ſame effect as ſpirit of vitriol; but ſpirit of nitre and aqua fortis excite no fulmination.

OF

OF THE

VOLCANO'S and EARTHQUAKES

IN

P E R U.

IT is a very eaſy matter to examine the internal diſpoſition of the earth in *Peru*; for the whole province is cut through with *Ravines* or great trenches, many of which are 200 toiſes or fathoms broad, and ſixty or eighty deep, and others twice as much. Some of them may probably have been the work of earthquakes, but the greateſt part are owing to rapid torrents of water which among the mountains in tempeſts are capable of carrying every thing before them, tho' at other times they are ſo reduced that one may frequently paſs them dry-ſhod. Sometimes the ſides of theſe trenches are cut quite perpendicular, and being purſued to their origin, appear to have been formed by a vertical fall of water.

'Tis only neceſſary to find out a place propes for deſcending down this kind of river beds, which ſeldom hold any quantity of water, in order to ſurvey and examine the qualities of the different *ſtrata* or layers of the earth. None of them diſcover any conſiderable marks of great inundations, ſo frequent in other countries. I have ſearched them with all poſſible care for ſea ſhells, but was never able to diſcover one. Probably the mountains of *Peru* are too high. There is a great quantity of that black ſand which the loadſtone attracts; it is eaſy to diſcern that theſe layers, whoſe colours are readily diſtinguiſhable, far from being the effect of repeated waſhings, are an expanſion of ſubſtances vomited out by volcano's; every thing ſeems to be the produce of fire. Some of theſe mountains are formed, to a certain depth, of mere cinders, pumice ſtones, and fragments of burnt ſtones of all ſizes, all which are ſometimes concealed under a bed of common earth, on which herbs and trees flouriſh. Theſe ſubſtances are diſpos'd in layers, of different thickneſs, diminiſhing as you recede from the mountain, to a foot, half a foot, an inch; but do not quite vaniſh in leſs than four or five leagues diſtance, till approaching another volcano, you begin to meet with them again.

All theſe particulars I remarked chiefly at the foot of the mountain *Cotopaxi*, which is now become a perfect truncate cone, having loſt its head. The baſe of this volcano has been made round and taken a regular form, from the rowling down of the ſeveral materials which were not thrown out with

with ſufficient force, or were of too light a nature to receive any great degree of motion. At the foot are beds of burnt ſtone reduc'd into ſmall parcels, five or ſix times a man's height in thickneſs. The thickeſt of them all being the uppermoſt; and I am very ſure that this extends alſo the fartheſt, and is hid under the good ſoil, which, 'tis likely, was at firſt nothing but aſhes. I am induced to believe that this upper bed of calcined ſtones is to be attributed to that terrible eruption which hiſtorians ſpeak of, after the death of *Atahualpa*, king of *Quito*, of which we have ſeen other extraordinary marks with the greateſt amazement; ſtones of eight or nine feet diameter, thrown to more than three leagues diſtance, ſeveral of which by the train they have formed, indicate plainly enough from what volcano they were projected. Theſe maſſy ſtones are no ways burnt, like thoſe which cover the foot of the mountain, nor could they have been thrown ſo far, but at the firſt effort of the exploſion; accordingly 'tis improbable that any like effect will hereafter happen, the mouth of the volcano being at this time 5 or 600 fathoms wide.

The *Indians* pretend that this accident had been foretold them, and that they look'd upon it as the fatal moment when it was in vain to defend themſelves longer againſt ſtrangers who were deſtin'd to ſubdue them, and had already made very great advances in their conqueſt: *Pedro Cieca de Leon*, *Garçilaſſo*, *Herrera*, and all the hiſtorians mention this: they attributed theſe predictions partly to *Huayana Capuc*, the twelfth and laſt emperor, father of *Atahualpa*; they called this mountain the volcano

O 2 of

of *Latacunga*, which is five or ſix leagues diſtant from it. If we may gueſs at its different eruptions by the number of the beds of burnt ſtones at its foot, without taking notice of ſome of the loweſt of them, which are broken and overturned, we muſt allow this conflagration to have been at leaſt the twentieth; but 'tis probable that each eruption ejects materials of different colours and kinds, and that they are thrown out ſucceſſively, according to their arrangement in the body of the mountain. However, it is paſt all doubt that it has raged ſeveral times, for the eruption of 1553 could not poſſibly furniſh all thoſe ſubſtances which are at this day viſible at the foot of this volcano. If all the beds had been elanc'd at the ſame time, the ſeveral ſettlements which the *Indians* had in that neighbourhood, ſome of which ſtill ſubſiſt, had been infallibly deſtroyed at once. But what epoch can we aſſign to thoſe overturned beds which we ſee below the reſt? Theſe had been ranged parallel like the other entire ones; but nature forgetting, as I may ſay, her gradual way of acting, threw this part of the *Cordiliere* into convulſions. I took particular notice of ſuch broken beds near a place called *Tioupoulou*, above four leagues from the volcano; they are above 40 feet deep: what a prodigious agitation muſt it have been that was able to break and tumble them one upon another as they now remain?

It was in all probability in times very remote, and moſt likely before the country was inhabited, that the vaſt maſs of pumice ſtones about ſeven leagues ſouth of *Cotopaxi* was formed. There are no no pumice ſtones to be found on the mountains, but of a moderate ſize, and all ſingle fragments: but here there are whole rocks of them, conſiſting of parallel banks each five or ſix feet thick, and covering more than a league ſquare, to what depth is unknown. Can one imagine what fire it muſt be that could put this enormous maſs in fuſion, and that all together at once and in the place where it now actually is? for it is manifeſt that it never was diſturbed, but ſettled cold on the very ſpot where it had been melted. The neighbouring parts have profited by this immenſe quarry, and the whole city of *Latacunga*, which has very fine houſes, is built out of it, ſince the earthquake which deſtroyed it in 1698.

The laſt conflagration of *Cotopaxi* in 1742, which happened before our eyes, did no miſchief, except by the melting of its ſnow; notwithſtanding that it opened a new mouth in its ſide about the middle of that part continually covered with ſnow, whilſt the flame conſtantly iſſued through the top of the truncate cone. There were two ſudden inundations, on the 24th of *June*, and the 9th of *December*, but the laſt was incomparably the greateſt. In the firſt place it muſt be noted that the water fell at leaſt 7 or 800 fathoms. The waves it formed in the valley were above ſixty feet high, and in ſome places it roſe more than 120 feet. Not to mention the infinite number of cattle which it ſwept away, it overturned 5 or 600 houſes, and deſtroyed 8 or 900 perſons. Theſe waters had 17 or 18 leagues to run, or rather to ravage, towards the ſouth of the *Cordiliere* before they could get all out of it at the foot of the mountain *Tongouragoua*; yet they took up no more than three hours in all that paſſage; which may afford ſome conception of their mean velocity, by which I would underſtand the mean between the prodigious rapidity they acquired at firſt by their fall, and their ſloweſt motion afterwards: and if we may judge from the ſeveral effects they produced at three or four leagues diſtance, they muſt have run 40 or 50 feet in a ſecond of time. Heavy ſtones of 10 or 12 feet diameter were removed 14 or 15 fathoms from their former places on a plain almoſt horizontal.

Every body at *Quito* was firmly of opinion, that the water iſſued from the inſide of the mountain, being led to think ſo, by a whimſical diſtinction of volcano's throughout all that country, into fiery and watery ones. It is not indeed impoſſible that waters ſhould be congeſted in the large cavities which are ſometimes formed in the upper parts of mountains, they may be ſupplied by the aſcending ſteam of the waters below, much in the manner which *Deſcartes* has explained. If the heat of the ſun be inſufficient, neighbouring ſubterraneous fires may furniſh a plentiful evaporation; and when the waters are collected above, it is not ſurprizing that they ſhould ſometimes bear down the walls or bounds of their confinement, and at once ſpread themſelves over the country. But no ſuch notion was conceived of what happened at *Cotopaxi*. To prove that the waters boiled in their baſin which was formed for them at the top of the mountain, and that it was the vehemence of this ebullition which

which threw them over the brims, they alledg'd the appearance of the dead corpſes below, which almoſt all looked as if they had been expoſed to the action of boiling water.

I got ſeveral particulars clear'd up to my ſatiſfaction by examining credible witneſſes on the ſpot. Many who eſcaped near the edge of the inundation aſſured me that the water was not in the leaſt hot. They perceived an oily matter which flamed and ſwam on the ſurface of the flood and was carried along in the front of it; and probably this was what affected the bodies in ſuch a manner. They likewiſe told me that when they heard a great noiſe, which the firſt fall in all likelihood occaſioned, the top of the mountain was ſurrounded with clouds; which abſolutely deſtroy'd the report of ſome who gave out that they ſaw as it were a river run over the brim of the volcano, like water running over the ſide of an inclined veſſel. It appeared to me at length after examining the extent of the ſpace it had covered, and all other circumſtances, that a very ſmall quantity of water might cauſe the whole diſaſter. In ſeveral parts the inundation did not continue a quarter of a minute. It was preceded by a deafening noiſe. They warned one another of the danger; but many, inſtead of running to elevated places, went rather to meet it. The water diſappeared in an inſtant; and one would have thought it had been a dream, but for the melancholy monuments it left behind it. I ſuppoſe that the ſnow towards the top of the volcano had been melted ſome time. That below being out of the influence of the fire retained its hardneſs, and formed a ſort of baſin with the outſide of the mountain. But the thaw continually encreaſing, the weight was too much to be ſupported, and ſo the water fell, and carried down with it large maſſes of ſnow, all reeking, which tho' broken by one another in their fall, meaſured ſome of them above 15 and 20 feet in thickneſs.

There was ſomething like this when a furious earthquake threw down the ſmall city of *Latacunga*, with a great many leſſer towns and villages as far as *Ambato*, which lie about the middle of our meridian. A very high mountain almoſt adjacent to the mountain *Chimboraco*, tumbled down, with ſeveral leſſer ones, upon which iſſued ſuch a great quantity of water as cauſed an inundation throughout the neighbourhood, if mouldering earth mixed with water into a mud may be ſo called; which mud however was ſo liquid as to run like brooks and rivers, whereof many marks ſtill remain. *Cargaviraco*, the higheſt of theſe mountains, has at this time but a moderate height. Others tumbled in part, one half falling, and the other remaining with ſuch a ſteep acclivity as renders them inacceſſible on that ſide. I had the curioſity to aſcend one of them called *Pugnalic*, I found an infinite number of crevices which compell'd me to walk with great caution, and the earth appeared extremely looſe. *Cargaviraco*, ſince it has loſt its height, has aſſumed the figure of a very flat cone, and muſt contain ſalts which promote congelation. Although it wants conſiderably of the height of the level which is taken for the loweſt limit of conſtant ſnow in the reſt of the mountains, yet its top is is covered with perpetual ſnow. It is very particular in this, that near it you ſee green fields planted with trees, which extend to the diſtance of ſome leagues from it. The fate of *Latacunga* was extremely deplorable. Whole families were buried under the ſame roof, and there was not a ſingle houſe that eſcaped without the death of ſomebody. This terrible ſcene was tranſacted on the 20th of *June* 1698, about an hour after midnight, and almoſt all the miſchief was done by the firſt ſhock.

It will not be ſurprizing that judicial aſtrology ſhould venture to prognoſticate earthquakes in *Peru*. The taſte of that vain ſcience prevails in all countries where true knowledge has not made any progreſs. A curious fellow who was deputy profeſſor of mathematics in the univerſity of *Lima*, publiſhed a work in 1729 with the title of *The Dial of Earthquakes*. At that time he was contented with barely pointing out the fatal hours in which there was reaſon to apprehend a ſtroke. But in 1734 he publiſhed another book containing a *Tragical Period* ſerving to diſtinguiſh the years ſubject to the ſame accidents; and he did not ſcruple to advance that if in 1729 his *dial* had been confirmed by 143 obſervations, he had now in 1734 collected 70 more equally conformable to it. It has been long ago remarked that maritime places are more expoſed to theſe terrible phænomena than inland countries. Caſt your eyes on all parts of the old world where there are any volcano's, and you will find them to be almoſt all ſituated in iſlands or near the ſea coaſt. It is not the *Alpes* for example, that are ſubject to earthquakes, but thoſe parts of

Italy

Italy which are the moſt advanced into the *Mediterranean.* The ſame holds good in *America.* It may ſometimes happen that ſtores of inflammable matters congeſted in the earth, want nothing but the mixture of water to take fire. But when the ſea riſes high, whether from the effect of the tides, or being ſimply accumulated by winds, it may waſh over into certain ſubterraneous canals, and ſo penetrate into many places which it could not any other ways reach.

From whence it manifeſtly follows that the ſeveral circumſtances of the moon's motion which produce any ſenſible effects with regard to the flux and reflux, may do the like alſo with regard to earthquakes, and the eruptions of volcano's. Thus an aſtrologer who is continually prattling about the dragon's head and tail, the moon's diſtance from the ſun, her ſituation in reſpect of her apogee and perigee, at the ſame time, delivering out every thing in a vague manner, as is their conſtant way, may chance to advance ſome particulars which will not ſeem abſolutely void of ſenſe. I cannot help thinking the ſubject worthy of a little conſideration: and will venture here in a few words to deliver the reſult of my own remarks, which come naturally enough into the plan of this relation.

The great number of particular cauſes which conduce to theſe terrible accidents, may poſſibly be one main reaſon that the concurrence of ſeveral ſuch cauſes, often ſupplies what is deficient on the part of others: but the particular inſtant of the effect in point of time, cannot but be very uncertain. The heat of the ſun may contribute a ſhare; at leaſt we

ſee

ſee that it promotes the inflammation of ſubſtances which chymiſtry inſtructs us to mix together, for repreſenting the conflagration of a volcano [a]. The city of *Lima* has been three times ruin'd, firſt in 1586, and, again in 1687, and in 1746. The firſt time the earthquake happened *July* the 9th, the two laſt in *October*, to wit the 19th and 28th, after the equinoctial tides might have introduced a great quantity of water into the ſubterranean caverns, and the ſun advancing into the ſouthern hemiſphere, had begun to heat it more. Three other earthquakes were beſides very conſiderable ones; that of *June* 17th 1678, which is no example to our remark, but the other two, that of 1630, and that of 1655, both fell out in *November*, to wit on the 27th and 13th.

So of the ſix great earthquakes which *Lima* has felt ſince its foundation, there are four of them which inſtead of being diſtributed indifferently through the ſeveral parts of the year, have happened in *October* and *November*. This particularity may perhaps be look'd upon as the effect of meer chance. But is it impoſſible that the return of the heat, and the great tides in *September* and *October*, might contribute thereto? The communication between the ſubterranean caverns may likewiſe be a means of the effect of the tides extending itſelf to a great diſtance. Among the ſeveral earthquakes which I felt myſelf, one of the moſt violent threw down ſome houſes near *Latacunga*, and killed ſeveral people. At the ſame time, tho' not preciſely at the ſame inſtant, cloſe

[a] See the tract immediately preceding this.

to

to a neighbouring mountain, a flame was ſeen to dart up through the water of a lake. This was in 1736, about the beginning of *December*. I have more obſervations of the like kind; and all things conſidered, it appears as fact to me, that tho' the *Peruvians* are expoſed to theſe dreadful phænomena at all ſeaſons, yet are they moſt ſubject to them in the laſt months of the year.

The author I was ſpeaking of, aſſerts that there is abſolutely no critical time except the ſix hours and ſome odd minutes that the moon is paſſing from the horary circle of 3 to that of 9; that is, the time of the reflux, for it is high water on almoſt all the coaſts of *America* in the *South Sea*, when the moon paſſes the horary circle of 3. But it ought to be well examined into how many different conditions muſt concur to make our author's rule exact. In the firſt place it is neceſſary that the focus of the conflagration ſhould be always in the ſame place, that the water ſhould follow the ſame rout, that it ſhould always penetrate with the ſame velocity, that the mixture ſhould take up preciſely the ſame time in its ignition. If theſe ſeveral conditions do not all take place at once, there muſt at leaſt be ſome ſort of compenſation to ſupply the defect. The earthquake which occaſion'd the deſtruction of *Lima* in 1746, happened when the moon, inſtead of paſſing from the horary circle of 3 to that of 9, was on the contrary, paſſing from that of 9 to that of 3. The author pretends that no danger is to be apprehended but when the moon's nodes are poſited in the malevolent ſigns of *Scorpio* or *Aquarius*: however at the

time

time of that disaster they were in the signs of *Virgo* and *Pisces*.

Scarce a week passes without some slight shocks and tremblings in *Peru*; if they are not felt in one place, they are in another. For the most part but little attention is given to them; and no body thinks it worth while to register them. An astrologer is therefore at full liberty to boast that the observation never contradicts his prognostick. It is the fatal earthquake alone that can bring his skill in question; but happily those are rare, and may besides happen as well at one time as at another. The precaution is commonly taken not to confine the prognostick within too narrow limits. Moreover the pretended rule can never fail of coinciding with some of the previous accidents or after consequences, and that is enough to save the wizzard's credit.

In a word, to proceed methodically, and discover, if there be in reality any such thing as a tragical period, a quite different road must be taken. We must begin with examining the most simple cases; and it seems that eruptions of volcano's should be the first object of observation. But whoever engages in this inquiry must expect to be puzzled with events extremely complicated. Earthquakes may be propagated by the bare contiguity of territories, even to an immense distance from the spot that is directly over the focus of conflagration. In every place are felt all the tremors which are excited round a certain point, and 'tis not to be known to what place they belong particularly; whereas volcano's are determin'd points, and con-

2 sequently

ſequently furniſh leſs equivocal obſervations. There is nothing regular in the return of their ragings. The ſame ſhould likewiſe hold good in regard to earthquakes, which for the reaſon juſt now aſſigned, ſhould be ſtill leſs confined to rules; ſince generally ſpeaking, they depend on a great number of caſualties for any particular place. Rain waters do without doubt very often produce the ſame effects as the waters of the ſea, and it ſhould be noted, that it is in the laſt months of the year that it rains the moſt in all the countries I have been ſpeaking of. Sometimes a very ſtrong tremor in the *Cordeliere* extends itſelf but over an inconſiderable ſpace. There is reaſon to imagine that the ſtock of the inflammable matter is then not very deep below the ſurface, and that the ſea has no ſhare in the accident, at leaſt no immediate one. The ſea contributes to many earthquakes, as well as the rain to ſeveral others; ſo that there is a two-fold cauſe of their frequency.

The compariſon of the eruptions of volcano's and earthquakes throws ſome light upon ſeveral particulars of theſe laſt. The volcano's when in a ſtate of high conflagration, act by fits; the flame and ſmoak are obſerved to iſſue out, almoſt always, by blaſts. When I was employed in one of our ſtations at *Senegualap*, my ſleep was diſturbed all night long by the bellowings of the volcano of *Mucas*, called *Sangai*. I was diſtant from it ſomething more than 18000 fathoms, yet the noiſe was horrible and awakened me every moment. This mountain is in the ſhape of a cone, whoſe ſides are perfectly ſtrait, and it wants only the

the vertex. All the neighbouring inhabitants are ſatisfied that the maſs of the mountain is continually decreaſing. Its preſent height above the level of the ſea is 2664 fathoms. The flame comes out from the top, and frequently a ſtream of melted matter runs down its ſides to the bottom. A *Ravine* of a foot broad has gotten the name of the *Sulphur River*. The bellowings of the volcano ſometimes form a claſhing noiſe like thunder, but they ſoon reſume their regular period, with a dull noiſe, with the repetition whereof I was ſo greatly incommoded. I obſerved likewiſe blaſts of ſmoak to iſſue out of *Cotopaxi* by equal intervals; there was about 42 or 43 ſeconds between each blaſt when I obſerved them. The ignited matter in the bowels of the volcano was doubtleſs dilated each time: but ſuch dilatation exhauſting it in part, the inflammation abated a little; which made room for the external air to enter anew, either by the opening at top, or by ſome other aperture. Perhaps alſo there might be at the ſame time an acceſſion of other inflammable matter, which found at that inſtant an eaſy admiſſion. Immediately the conflagration acquired a new force which produced a freſh iſſue of ſmoak or another bellowing.

The matters which take fire in the bowels of the earth and cauſe earthquakes, muſt be ſubject to the ſame alternatives. When the fire is kindled up in an hollow cavern, the dilatation of the inflamed matter and of the air muſt be extended very far and act in other ſubterraneous hollows which have a communication with the former. The ceiling of the vault is puſhed upwards with great

great force, and it may be alſo puſhed laterally tho' the ſtock of the materials be exactly under. The direction of the effort depends then upon the horizontal ſituation or the inclination of the vault; and this is the cauſe that ſometimes the walls of houſes are, or are not ſpared according to the way they happen to be ſituated. The ceiling of the vault returns to its former place by repeated oſcillations which are independent of the action of the fire; the effort of the exploſion ceaſing a little, at the ſame time that the air is over much compreſs'd in all the neighbouring caverns, whence a violent reflux towards the place of the conflagration, and a new fit and a ſtronger ſhock; and thus are brought about the reiterations before mentioned, whoſe intervals muſt be ſenſibly equal, till ſome very conſiderable alteration happens either in the ſubterraneous diſpoſition or in the inflamed materials. The feebleſt ſhocks are thoſe of a ſoil once ſhaken, the ſtrongeſt are thoſe that are the immediate effect of an inflammation; which are analogous to the bellowings of volcano's, and muſt be repeated with more or leſs frequency, according to the facility with which the matters are ignited, and likewiſe according to the proportion of their bulk to the extent of the ſpaces within which they exert their force.

THE

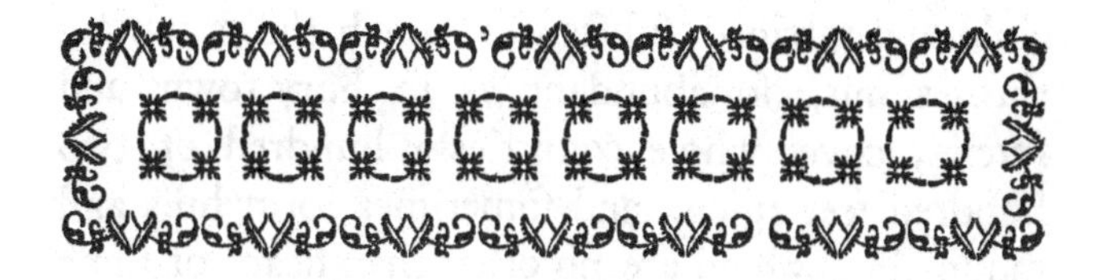

THE

NATURAL HISTORY

OF

EARTHQUAKES and VOLCANO's.

BUrning mountains, called volcano's, contain within them ſulphur, bitumen and other materials which are the *pabulum* of a ſubterraneous fire, whoſe effect, more violent than that of gun-powder or thunder, has been aſtoniſhing in all ages, terrified mankind, and laid the earth deſolate. A volcano is a cannon of an immenſe ſize, whoſe aperture is often more than half a league in circumference. Out of this vaſt mouth are vomited torrents of ſmoak and flames, rivers of bitumen, ſulphur and melted metal, clouds of aſhes and ſtones, and ſometimes it ejects enormous maſſes of rocks to ſeveral leagues diſtance, ſuch as no combined human ſtrength could be capable of putting in motion. The conflagration is ſo horrible, and the quantity of burning, melted, calcin'd and vitrified ſubſtances which the mountain throws out, ſo abundant as to bury towns and foreſts, cover whole countries a hundred or two hundred feet thick, and ſometimes form hills and mountains, which are no other than heaps of thoſe compacted matters. The action of the fire is ſo vehement, and the force of the exploſion ſo powerful, as by its reaction to produce ſhocks ſufficient to ſet the earth in a tremor, agitate the ſea, overthrow mountains, deſtroy cities and the moſt ſolid edifices, and that to very conſiderable diſtances.

Theſe effects, though natural, have been lookt upon as prodigies, and notwithſtanding we behold in miniature, effects of fire pretty ſimilar to thoſe of volcano's; yet the grand, of what nature ſoever it be, has ſo irreſiſtible a power of amazing, that I am not ſurprized ſome writers have taken theſe mountains for ſpiracles of central fire, and the vulgar for the mouths of hell. Aſtoniſhment begets fear, and fear generates ſuperſtition. The inhabitants of the iſle of *Iceland* do believe the bellowings of their volcano to be the cries of the damned, and that it's eruptions are the effects of the fury and deſpair of its wretched priſoners.

All this however is no more than noiſe, fire and ſmoak. There are in mountains veins of ſulphur, bitumen and other inflammable materials, and at the ſame time there are minerals, as pyrites, capable of fermenting, and which this in reality does whenever it is expos'd to air or moiſture; it abounds every where in vaſt quantities, kindles and produces an exploſion in proportion to the quantity of the inflamed ſubſtances, the effects of

 which

which are greater or leſs in the ſame proportion: ſuch is the idea of a volcano in the mind of a naturaliſt, who may eaſily imitate the nature of thoſe ſubterranean fires, by mixing together a certain quantity of ſulphur and filings of iron, and burying them under ground. Thus will a ſmall volcano be produced, whoſe effects are the ſame, regard being had to proportion, as thoſe of great ones, for it ignites by mere fermentation, throws off the earth and ſtones which cover it, ſmoaks, flames and explodes.

In *Europe* there are three noted volcano's, *Ætna* in *Sicily*, *Hecla* in *Iceland*, and *Veſuvius* near *Naples* in *Italy*. *Ætna* has burnt time immemorial, its eruptions are very violent, and the ſubſtances it throws out ſo copious, that you may dig in them to the depth of 68 feet, where have been found pavements of marble, and the remains of an ancient city which was covered and buried under that prodigious bed of ejected earth, after the like manner as the city of *Heraclea* was covered by matters thrown out of *Veſuvius*. New fiery mouths were formed in *Ætna* in 1650, 1669, and at other times: the flame and ſmoak of this volcano may be ſeen as far as *Malta*, which is 60 leagues; ſmoak is continually ariſing out of it, and at certain times it vomits out flames and variety of different ſubſtances with great impetuoſity. In 1537 there was an eruption of this volcano which occaſioned an earthquake throughout all *Sicily* for twelve days, and overthrew a great number of houſes and edifices; it ceaſed by the opening of a new mouth of fire which burnt up every thing within five

 leagues

leagues of the mountain. Aſhes were thrown out in ſuch abundance that they were carried even into *Italy*, and ſhips at a very great diſtance from the *Sicilian* ſhore were incommoded by them.

This volcano has at preſent two principal mouths, one narrower than the other; theſe two openings always ſmoak, but no fire is perceived except in the times of the eruption: it is ſaid that ſtones have been projected out of it to the diſtance of 60,000 paces,

In 1693, there happened a terrible earthquake in *Sicily* occaſioned by a violent eruption of the volcano, which entirely deſtroyed the city of *Catanea*, and killed above 60,000 perſons in that place only, beſides great numbers in the neighbouring towns and villages.

Hecla ſhoots forth its fires through the ice and ſnow of a frozen ſoil; and yet its eruptions are no leſs violent than thoſe of *Ætna*, and other volcano's of the more ſouthern climes. It throws out vaſt quantities of aſhes and pumice ſtones, and at ſome times boiling water; there is no dwelling within ſix leagues of this volcano. The whole iſle of *Iceland* abounds in ſulphur. The hiſtory of its moſt violent eruptions may be found in a book written by *Dithmar Bleffken*.

Mount *Veſuvius*, according to the account of hiſtorians, has not always burned, nor did it begin to do ſo before the ſeventh conſulate of *Titus Veſpaſian* and *Flavius Domitian* [a]. As ſoon as the ſummit

[a] It is however a point not ſettled among the learned, whether this great eruption was the firſt of that nature, or if

mit was opened the volcano threw out ſtones and rocks, and afterwards fire and flames in ſuch abundance that they burnt two neighbouring cities, and ſo thick a ſmoak that it darkened the light of the ſun. *Pliny* the elder ventured to take too near a view of it, and was ſuffocated with its fumes [b]. *Dion Caſſius* relates that this eruption of *Veſuvius* was ſo violent as to throw out aſhes and ſmoak with that violence as to carry them to *Rome*, and even acroſs the *Mediterranean* into *Egypt*. One of the two cities that were overwhelmed with the rejected matter of its firſt conflagration was *Heraclea*, rediſcovered of late years at 60 feet depth under the ſaid matter, whoſe ſurface in proceſs of time was become arable, and accordingly cultivated. The relation of the diſcovery of *Heraclea* is in every ones hands, it were only to be deſired that ſome body well verſed in natural hiſtory, would be at the pains of carefully examining the ſeveral ſubſtances which compoſe this immenſe thickneſs, and at the ſame time note the diſpoſition and ſituation of them, the alterations that they have produced, or ſuffered themſelves, the direction which they followed, and the degree of hardneſs they have acquired, &c.

if ſomewhat of the like kind had not happened in ancient ages. M. L'Abbé *Bannier* has taken ſome pains about this particular, and has found in *Strabo* and *Diodorus Siculus*, that there is mention of very ancient veſtiges of the flames of *Veſuvius*. To theſe the Abbé adds the authority of ſeveral poets, and upon the whole, concludes that there had been fiery eruptions from that mountain in very remote times. *J. B.*

[b] See the younger *Pliny*'s epiſtle to *Tacitus*.

 There

There is ſome ground to believe that *Naples* is ſituated on a hollow bed of roaſted minerals, ſeeing *Veſuvius* and the *Solfatara* do appear to have internal communications. For when *Veſuvius* burns, the *Solfatara* throws out flames, and when that ceaſes, the *Solfatara* does ſo too. The city is ſituated nearly at an equal diſtance between them.

One of the laſt and moſt violent eruptions of *Veſuvius*, was that of the year 1737, when the mountain vomited a large torrent of red hot and melted metalline ſubſtances through ſeveral mouths, which ſpread over the country, and made its way even into the ſea. M. *de Montealegro*, who communicated the relation to the academy of ſciences, ſaw with horrour one of theſe rivers of fire, and obſerved that its courſe was ſix or ſeven miles from its ſource to the ſea, its breadth being 50 or 60 paces, its depth 25 or 30 palms, and in ſome hollows of the valleys, more than 120 palms. The matter as it roll'd along look'd like a ſkum which runs out of the furnace of a forge, *&c.* [c].

In *Aſia*, more eſpecially in the iſlands of the *Indian* ocean, there is a great number of volcano's, one of the moſt famous of which is mount *Albours*, near mount *Taurus*, eight leagues from *Herat*. Its top is continually ſmoaking, and it frequently throws out flames and other ſubſtances ſo abundantly, that the whole country round is covered with them. In the iſland of *Ternate* there is a volcano, which ejects a ſubſtance like pumice ſtone in immenſe quantities. Some travellers affirm that this volcano burns more furiouſly about the time

[c] *Hiſt. de l'Acad. ann.* 1737. p. 7 and 8.

of

of the equinoxes, than in other ſeaſons of the year, becauſe certain winds do then blow which contribute to ignite the matter which has ſo many years nouriſhed its fires[d]. The iſle of *Ternate* is but ſeven leagues round, being no other than the ſummit of a mountain. From the ſhore you aſcend every way towards the middle of the iſland where the volcano is elevated to a very conſiderable height, and is in a manner inacceſſible. It furniſhes ſeveral ſprings of freſh water which run down its ſides; and when the air is calm, and the ſeaſon mild, the gulph is in a leſs agitation than when the winds are violent[e]. This proves that the fire of volcano's does not come from any great depth within the mountain, but from its upper part, or at leaſt, not far down, and that the focus of the conflagration cannot be a great way from the top; for if it were not ſo, great winds could not contribute to their rage. There are ſome other volcano's among the *Molucca* iſlands. In one of the *Mauritian* iſlands, about 20 leagues from the *Molucca's*, there is a volcano as violent in its effects as that of *Ternate*. The iſland of *Sorca*, one of the *Molucca's*, was once inhabited; in the middle of it was a volcano, being a very high mountain. In 1693 this volcano vomited out bitumen and other inflamed ſubſtances, in ſo great a quantity as to form a burning lake, which extended by degrees till it entirely covered the whole iſland[f]. In *Japan* are alſo ſeveral volcano's; and in the neighbouring iſles navigators have taken notice of many mountains

[d] Voyages *d'Argenſola*, tom. i. p. 21. Voyage de *Schouten*. [f] Philoſ. Tranſact. abridg'd, vol. ii. p. 391.

 whoſe

whoſe tops caſt up flames in the night and ſmoak in the day. There are alſo ſeveral burning mountains in the *Philippine* iſlands. One of the moſt famous volcano's of the *Indian* ocean, and at the ſame time one of the neweſt, is near the town of *Panarucan* in the iſland of *Java*. It opened in 1586, and there is no account of its having ever burned before that time. In its firſt eruption it diſcharged an immenſe quantity of ſulphur, bitumen and ſtones. The ſame year the mountain *Gounapi* in the iſland of *Banda* (whoſe laſt conflagration was not above 17 years ago) opened with a moſt terrible noiſe, and vomited out rocks and ſubſtances of every kind. Beſides all theſe there are other volcano's in the *Indies*, as in *Sumatra*, and in the northern part of *Aſia*, beyond the river *Jéniſcea*, and the river *Péſida*, but theſe two laſt are not very well known.

In *Africa* there is a mountain, or more properly a cavern, called *Beni-guazeval*, near *Fez*, which always caſts forth ſmoak, and ſometimes flames. One of the *Cape de Verd* iſlands, called *Fuego*, is one huge mountain which burns inceſſantly; this like the reſt throws out much aſhes and ſtones, and the *Portugueze* who have ſeveral times attempted to ſettle inhabitants in the iſland, have been obliged to drop their project, for fear of the effects of the volcano. In the *Canaries* the pike of *Teneriffe* which paſſes for one of the higheſt mountains upon earth, throws forth fire, aſhes and great ſtones; from its top run down rivulets of melted ſulphur on the ſouth ſide, through thick beds of ſnow, which by ſoon coagulating, forms veins that may be ſeen at a great diſtance. In

In *America* there is a great number of volcano's, eſpecially in the mountains of *Peru* and *Mexico:* That of *Arequipa* is one of the moſt famous; it oftentimes occaſions earthquakes, which are more frequent in *Peru*, than in any part of the known world. The volcano of *Carapa*, and that of *Malaballo*, are according to the relation of travellers, the moſt conſiderable after that of *Arequipa*. But there are a great many others of which we have no very exact knowledge.

In *Mexico* are divers volcano's, the moſt conſiderable of which are *Popochampeche* and *Popocatepac*, near which latter *Cortez* march'd to *Mexico*, and ſome of his *Spaniards* aſcended to the top and found the mouth of it half a league round. Sulphureous mountains have alſo been found in *Guadeloupa*, *Tercera*, and others of the *Azores* iſlands; and if all the mountains from whence flame or ſmoak ariſes, are to be ranked among volcano's, above 60 of them may be reckoned up; thoſe we have ſaid the moſt of are the remarkable ones, ſuch as will endure no inhabitants about them, and which project ſtones and minerals to a mighty diſtance.

The numerous volcano's among the *Cordelieres*, as I have obſerved, are the occaſion of frequent, and almoſt continual earthquakes, ſo that no ſtone buildings in that country are carried higher than the firſt floor, whatſoever is added above, is of light wood and ruſhes. In ſome of theſe high mountains are found many precipices and large openings, whoſe ſides look black and burnt, as does the precipice of mount *Ararat* in *Armenia*, called

 the

the *Abyss*; these abysses are the mouths of ancient volcano's, now in a state of extinction.

Of late years there happened an earthquake at *Lima*, the effects whereof were most terrible; the city of *Lima* and the port of *Callao* were almost totally overwhelmed by it. The sea covered every edifice with its waves, one tower alone excepted, so that all the inhabitants were drowned: of 25 ships which were at that time in the port, four were carried a league in land; the rest the sea swallowed up. Of the great city of *Lima* there remained only 27 houses standing, multitudes of persons were crushed to death, especially monks and nuns, their buildings being lofty and of solid materials. This disaster happened in the night time in the month of *October* 1746, the shock having lasted a quarter of an hour.

Near the port of *Pisco* in *Peru*, there was formerly a famous city situate on the sea coast, but it was intirely ruin'd and laid waste by the earthquake of the 19th of *October* 1682: for the sea having exceeded its wonted bounds washed it quite away with all its inhabitants.

If we consult historians and travellers, we shall meet with accounts of several earthquakes and eruptions of volcano's, whose effects have been no less terrible than those I have related. *Possidonius*, as cited by *Strabo*[g], relates that there was a city in *Phenicia*, situated near *Sidon*, which was swallowed up by an earthquake; and with it the neighbouring territory, and two thirds of the said city of *Sidon*, and that this effect did not take place sud-

[g] Lib. i.

denly,

denly, but moſt of the inhabitants had time enough to eſcape: that this earthquake extended itſelf almoſt over all *Syria*, and even to the *Cyclades* iſlands, and to *Eubea*, where the fountains of *Arethuſa* ſtopp'd all at once, and flow'd not again till ſeveral days after, and then by new apertures at a conſiderable diſtance from the old ones; and that the earthquake did not give over ſhaking the iſland in one place or other, till the earth had opened in the valley of *Lepanta*, and thrown out abundance of ignited matter. *Pliny* relates [h] that in the reign of *Tiberius* there happened an earthquake which demoliſhed 12 cities of *Aſia*, and in another place [i] he ſpeaks of a prodigy occaſioned by an earthquake in the following terms: *Factum eſt ſemel (quod equidem in Etruſcæ diſciplinæ voluminibus inveni) ingens terrarum portentum, Lucio Marco, Sex. Julio Coſſ. in agro Mutinenſi. Namque montes duo inter ſe concurrerunt crepitu maximo adſultantes, recedenteſque, inter eos flamma fumoque in cœlum exeunte interdiu; ſpectante e via Æmilia magnâ equitum Romanorum, familiarumque et viatorum multitudine. Eo concurſu villæ omnes eliſæ, animalia permulta, quæ intro fuerant, exanimata ſunt*, &c. St. *Auſtin* ſays [k] that by a great earthquake a 100 towns were overthrown in *Lybia*. In the days of *Trajan* the city of *Antioch*, and a great part of the adjacent country was ſwallowed up by an earthquake; and in the time of *Juſtinian*, in 528, that city was a ſecond time deſtroyed by the ſame cauſe, with above 40,000 of its inhabitants; and 60 years after that, in the time of St. *Gregory*, it was viſited by a third earth-

[h] Lib. i. [i] *Ibid.* [k] Lib. ii. *de Miraculis*. cap. 3.

quake,

quake, with the loſs of 60,000 inhabitants. In the reign of *Saladin*, in 1182, moſt of the cities of *Syria* and of the kingdom of *Jeruſalem* were deſtroyed by the ſame cauſe. In *Apulia* and *Calabria*, earthquakes have been more frequent than in any other part of *Europe*. In the pontificate of *Pius* II. all the churches and palaces of *Naples* were thrown down, near 30,000 perſons killed, and the inhabitants that remained alive were forced to live in tents till they could get their houſes rebuilt. In 1629 there were earthquakes in *Apulia* which deſtroyed 7000 perſons; and in 1638 the city of St. *Euphemia* was ſwallowed up, and a ſtinking lake left in its place; *Raguſa* and *Smyrna* were likewiſe almoſt deſtroyed. In 1692 an earthquake extended over *England*, *Holland*, *Flanders*, *Germany* and *France*, but was felt moſt ſenſibly along the ſea coaſts, and near great rivers: it ſhook at leaſt 2600 ſquare leagues, yet it laſted but two minutes, and the motion was more conſiderable on mountains than in valleys[1]. In 1688 on the 10th of *July*, there was an earthquake at *Smyrna*, which began with a motion from weſt to eaſt. The caſtle fell firſt, its four walls opening and ſinking ſix feet into the ſea: this caſtle, which was an iſthmus, is now a real iſland a 100 paces from the land. The walls which ſtood eaſt and weſt are fallen, thoſe that ſtood north and ſouth ſtill remain. The city, which is ten miles from the caſtle, was thrown down preſently after; there were in ſeveral places openings of the earth, from whence ſubterraneous noiſes were heard; before

[1] *Ray's* diſcourſes, p. 272.

night

night five or ſix ſhocks were felt, the firſt laſted about half a minute. The roads were agitated, the ground in the city ſunk two feet, not above a fourth part of the buildings ſtood, and thoſe chiefly were founded on rocks; they reckon that 15 or 20,000 perſons were loſt [m]. In 1695 in an earthquake which was felt at *Bologna* in *Italy*, it was particularly remarked, that the waters were troubled the day before [n].

There was a great earthquake at *Tercera* on the fourth of *May* 1614, which in the city of *Angra* overthrew eleven churches and nine chapels, beſides private houſes; and in the city of *Praya* it was ſo terrible, that ſcarce a houſe was left ſtanding; and on the 16th of *June* 1628, happened a horrible earthquake in the iſland of St. *Michael*, near the land the ſea opened, and an iſland aroſe in a place over which there was before 150 fathoms of water, which iſland was a league and an half long, and above 60 fathoms high [o].

There was another earthquake in 1591 which began the 26th of *July*, and laſted in the iſland of St. *Michael* till the 12th of the following month: *Tercera* and *Fayal* were ſhaken the next day with ſuch violence, that they ſeemed as though they were turned about, however theſe dreadful ſhocks were repeated there but four times, whereas in St. *Michael* they ceaſed not a moment for 15 days: the iſlanders having abandoned their houſes, which drop'd as they left them, were all that while expoſed to the injuries of the air. A whole city

[m] *Hiſt. de l'Acad. des ſciences, ann.* 1688. [n] *Hiſt. de l'Acad. ann.* 1696. [o] *Mandelſo*'s voyages.

called

called *Villa Franca*, was overturned to its foundations, and most of the inhabitants crushed under the ruins. In many places the plains rose up into hills, and in some the mountains sunk or changed their situation. From out of the ground issued a fountain of fresh water, which run four days, and then was dried up at once: besides this there was so violent an agitation in the sea and air, that the horrid sound of it resembled the bellowings of forests of savage beasts; many died of fear. There were no vessels in the harbours which did not undergo the utmost danger; and others which were at anchor, and some under sail 20 leagues off these islands, were yet more roughly dealt with. Earthquakes are common in the *Azores*; 20 years before one happened in the island of St. *Michael*, which overset a very high mountain[p]. In the month of *September* 1627, at *Manilla*, an earthquake levell'd one of the two mountains called *Carvallos*, in the province of *Cagayan*; in 1645 the third part of the city was ruined by a like accident, and 300 people perished; the next year, it suffered by another: the old *Indians* say, they were heretofore still more terrible; for which reason they built their houses of wood only; as the *Spaniards* do now above the first story. The number of volcano's in that island confirm what has been said. For at certain times they vomit out flames, shake the earth, and work the several effects which *Pliny* ascribes to those of *Italy*; that is, to shift the beds of rivers, cause the neighbouring seas to retreat, fill all places about them with ashes, and project great stones to

[p] Gen. Hist. of Voyages vol. i. p. 325.

a

a vaſt diſtance with a noiſe lowder than that of ordnance [q].

In the year 1646 the mountain of the iſland of *Machian* was ſplit aſunder with dreadful cracks and noiſes, by an earthquake, an accident not rare in that country; ſuch a quantity of fire iſſued out of the rent, as conſumed ſeveral negro plantations with their inhabitants: this prodigious aperture was to be ſeen in 1685, and 'tis very probable that it ſtill ſubſiſts, it was called the *Wheel-rut* of *Machian*, becauſe it ran from the top down to the bottom of the mountain like a hollow way [r].

The hiſtory of the *Pariſian* academy mentions the earthquakes of *Italy* in 1702 and 1703, in the following manner: the earthquake began in *Italy* in *October* 1702, and continued till *July* 1703; the parts which ſuffered moſt, as alſo where they began, are the city of *Norcia* with its dependencies, in the *Eccleſiaſtical State*, and the province of *Abruzzio*: theſe countries are contiguous and ſituated at the foot of the *Apennine*, on the ſouth ſide.

Theſe earthquakes were frequently accompanied with frightful noiſes in the air, and the ſame noiſes have alſo often been heard without any earthquake, the ſky being very ſerene. The earthquake of *February* 2, 1703, the moſt violent of them all, was accompanied, at *Rome* at leaſt, with very ſerene weather and calm air; it laſted there half a minute, but at *Aquila*, the capital of the *Abruzzio*, three hours. It deſtroyed the whole city of *Aquila*,

[q] Voyage *de Gimelli Careri*, p. 129. [r] Conqueſt of the *Moluccas*, vol. iii. p. 318.

buried

buried 5000 perſons in the ruins, and committed great ravage round about.

Commonly the oſcillations of the earth were from north to ſouth, or nearly ſo, which was diſcovered by the vibrating of the chandeliers in churches.

Two openings were made in a field, out of which were thrown a quantity of ſtones with violence, which covered it all over and render'd it barren; after the ſtones it threw out, from the ſame openings, two ſpouts of water a great deal higher than the tops of the talleſt trees, which laſted a quarter of an hour, and inundated all quite to the neighbouring countries: the water was white, like ſoap-ſuds, and without any taſte.

A mountain near *Sigillo*, a village about 22 miles from *Aquila*, had upon its top a pretty large plain invironed with rocks which were as a wall to it. The earthquake of the ſecond of *February* changed that plain into a gulph of unequal breadth, its greateſt diameter being 25 fathoms, and its leaſt 20: the depth of it cannot be meaſured, and has been found to exceed 300 fathoms. At the time this opening was formed, flames were obſerved to iſſue out, and after them a very thick ſmoak which laſted three days with ſome interruptions.

At *Genoa* on the firſt and ſecond of *July*, they had two ſmall tremors, the laſt only felt by people on the mole. At the ſame time the ſea in the port ſunk ſix feet, ſo that the galleys touched ground, and this ſhallow laſted near a quarter of an hour.

The

The ſulphury water in the road from *Rome* to *Tivoli*, diminiſhed two feet and a half in depth, both in the baſin and the canal. In ſeveral places of the plain called *Teſtine*, there were ſprings and brooks which had made it all marſhy, but now it is perfectly dry. The water of the lake called *Enfer*, likewiſe diminiſhed three feet in depth: in the place of ancient ſprings now dried up, new ones have burſt out about a mile from the former, ſo that in all probability they are the ſame waters, which have alter'd their courſe [s].

The earthquake which formed the *Monte di Cenere* near *Puzzoli* in 1538, filled the *Lucrine* lake at the ſame time with ſtones, earth and aſhes, ſo that the lake is now a marſhy ſoil [t].

Some earthquakes are felt a great diſtance at ſea, Dr. *Shaw* relates [u] that in 1724, being on board the *Gazelle*, an *Algerine* ſhip of 50 guns, they felt ſuch violent ſhocks one after another, as if the weight of 20 or 30 tons had been let fall from a good height on the ballaſt. This was in a part of the *Mediterranean* where they had more than 200 fathom water: he adds that others had felt much more conſiderable earthquakes in other places, and one among the reſt 40 leagues to the weſt of *Liſbon*.

Schouten[w], ſpeaking of an earthquake which happened in the *Molucca's*, ſays, that the mountains were ſhaken, and ſhips that were at anchor in 30 or 40 fathom water, were jerked as if they had ran aſhore, or came foul of rocks; that daily experience ſhews that the ſame thing happens in the o-

[s] *Hiſt. de l'Acad. ann.* 1704. p. 10. [t] *Ray's* Diſcourſes, p. 12. [u] Travels, p. 303. [w] Tom. vi. p. 103.

cean where no bottom can be found, and that in earthquakes veſſels are violently toſs'd on a ſudden though the ſea be perfectly calm.

Le Gentil [x] ſpeaks of earthquakes whereof himſelf was witneſs, in the following terms. " I have " made ſome remarks on earthquakes; firſt, that " half an hour before the tremor, all animals ſeem " frightned, horſes neigh, break their halters, and " run out of the ſtables, birds are ſtunned as it " were, and come in a doors, rats and mice come " out of their holes, *&c.* Secondly, that ſhips at " anchor ſuffer ſuch violent agitations, as to ſeem " to be falling aſunder, their guns break looſe, " and their maſts ſpring; this is more than I could " have eaſily believed, had not many unanimous " teſtimonies convinced me. I know well that the " bottom of the ſea is a continuation of the land; " that if this land be ſhaken, it communicates the " ſhock to the waters it ſuſtains; but the thing " which I cannot form a conception of, is that irregular motion of a ſhip whereof all its ſeveral " parts do participate, as if the whole veſſel were " a part of the earth, and did not ſwim in a fluid; " whereas I ſhould think ſhe ſhould be liable to no " other motions than thoſe ſhe experiences in a " ſtorm: beſides, on the occaſion I am ſpeaking " of, the ſurface of the ſea was ſmooth, almoſt " without a wave, and the whole agitation muſt " be wholly internal, as the wind could have no " concern in the earthquake. Thirdly, that if the " cavern of the earth wherein the ſubterranean fire " is confined, runs north and ſouth, and if a city

[x] *Nouveau voyage autour du Monde, tom.* i. *p.* 172, &c.

" over

"over it be situate in a parallel direction thereto, "all the houses will be overthrown; whereas if the "same vein or cavern crosses the town, the da-"mage will be considerably less."

It happens in countries subject to earthquakes, that whenever a new volcano is formed, the earthquakes cease, and are no more sensibly felt, but in violent eruptions of the volcano, as was observed in the island of St. *Christopher*[y].

The excessive ravages occasioned by earthquakes have induced some naturalists to imagine that the mountains and other inequalities on the surface of the globe, are the mere effects of subterraneous fires, and that all the irregularities we discern over the whole earth, are to be attributed to the violent shocks and subversions which they have produced: *Ray*, for instance, is of this opinion; he believes that all mountains have been formed by earthquakes, or explosions of volcano's, as the *Monte di Cenere*, the new island near *Santorini*, &c. but he has not taken due notice, that the small elevations formed by the eruption of a volcano, or by the action of an earthquake, are not inwardly composed of horizontal strata, as all other mountains are, for by digging into *Monte di Cenere*, there are found calcined stones, pumice stones, ashes, burnt earth and dross of iron, all mingled together like a heap of rubbish. Besides if the great mountains of the earth, as the *Cordilieres*, *Taurus*, the *Alpes*, &c. had been produced by earthquakes and subterraneous fires, the prodigious force requisite to raise those enormous masses, must at the same time

[y] *Philos. Trans.* abridged, vol. ii. p. 392.

have destroyed a good part of the surface of the globe, and the effect of the earthquake would have been extremely, nay inconceivably violent, since the most extraordinary earthquakes recorded in history, have not had force enough to raise mountains. There was one, for example, as *Ammianus Marcellinus* reports [z], in the days of *Valentinian* the first, which was felt all over the known world, but it is not said, great as it was, to have raised one mountain.

It must however be own'd that it will appear from calculation, that though an earthquake may be powerful enough to raise a mountain, yet it would not be sufficient to displace the rest of the globe.

For let us suppose for a moment, that the chain of high mountains which traverses *South America* from the point of *Terra Magellanica* to the mountains of *New Grenada* and the *Gulph of Darien*, had been raised all at once by an earthquake, and then let us compute the effect of this explosion. This chain is about 1700 leagues long, and at a mean about 40 leagues broad, including the *Sierras*, or mountains of less elevation than the *Andes:* the surface is about 68,000 square leagues: I suppose the thickness of the matter displac'd by the earthquake to be one league, or that the mean height of these mountains, from the top to the bottom, or rather indeed to the caverns, which in this hypothesis must support them, is but a league, which will be easily granted; then, I say, the force of the explosion or earthquake will have elevated to the height of

[z] Lib. xxvi. cap. 14.

a league

a league a quantity of earth equal to 68,000 cubic leagues: but, action being equal to reaction, this explosion will have communicated to the whole globe, the same quantity of motion: now the whole globe is 12,310,523,801 cubic leagues; from whence substracting 68,000 there remains 12,310,455,801 cubic leagues, whose quantity of motion is equal to that of 68,000 cubic leagues raised one league; whence it appears that the force requisite to have displaced 68,000 cubic leagues, and remove them one league, would not have displaced the rest of the globe a single inch.

There would then be no absolute impossibility that the mountains have been raised by earthquakes, if their internal composition, as well as their external form were not evidently the work of the waters of the sea. The internal is composed of regular and parallel beds, filled with sea shells; the external of a figure whose angles every where correspond; is it credible that so uniform a composition and so regular a form should be produced by irregular shocks and sudden explosions?

But as this opinion has prevailed with several naturalists, and as it seems to me that the nature and effects of earthquakes are not clearly understood, I esteem it necessary to advance some ideas which may serve to throw light on the subject.

The earth having undergone great alterations on its surface, there are even to very considerable depths, holes, caverns, subterraneous rivulets and empty spaces, which sometimes have communications one with another by chinks and guts. Of caverns there are two kinds, the first is produced by the action of subterraneous fires and volcano's; the action of the fire lifts up, shakes and disperses to a distance whatever matters are over it, and at the same time rends and disranges those of either side of it, and so forms caverns, grottos, hollows and irregular dens, but these seldom occur but on round high mountains that have volcano's, and this species of caverns produced by the action of fire, are rarer than the caverns of the second kind, which are produced by waters. We have seen that the different strata of which superficial parts of the terrestrial globe consists, are all interrupted by perpendicular fissures of which I shall explain the origin hereafter; the waters of rain and vapours, descending by these, are collected together upon clay, and form springs and brooks; by their natural motions they find out all small cavities and vacuities, and have a constant tendency to form themselves passages, till they procure some egress; carrying along with them at the same time sand, earth, gravel and other substances which they are capable of comminuting, and so gradually, as I may say, paving themselves ways, and forming a kind of little channels or trenches; at length they run out, either on the surface of the earth or into the sea, in the form of springs: the matters they carry off with them leave vacuities, whose extent may probably be very considerable, and these vacuities form grottos and caverns, whose origin, it appears, is very different from that of the caverns produced by earthquakes.

Earthquakes are of two kinds; one of them is occasioned by the action of subterraneous fires and explosions of volcano's, and these are felt but to

small

ſmall diſtances, and at the time the volcano's are raging, or before their firſt eruption. When the materials which conſtitute ſubterraneous fires begin to ferment, wax hot, and break out into flame, the fire exerts itſelf *quaquaverſum*, or in every direction; and if it cannot naturally meet with vents, it raiſes the earth and procures itſelf a paſſage by diſperſing it, and thus produces a volcano, whoſe effects are reiterated, and ſubſiſt in proportion to the quantity of the inflammable materials. If the ſhock be conſiderable, a ſuccuſſion and ſlight commotion may be all the conſequence, at moſt a gentle earthquake, without the eruption of any volcano. The air generated and rarefied by the ſubterraneous fire, may likewiſe find out ſmall apertures to eſcape at, in which caſe again, the utmoſt conſequence will be no more than an earthquake without any eruption or volcano: but when the ignited matter is congeſted in abundance, and pent up by ſolid and compact ſubſtances, a commotion and a volcano will be the conſequence. Now theſe ſeveral commotions make but the firſt ſpecies of earthquakes, and can ſhake no very great ſpace. A very violent eruption of *Ætna*, for example, may excite a tremor all over *Sicily*, but will never extend to 3 or 400 leagues. When any new mouths of fire happen to open in *Veſuvius*, tremors are felt in its neighbourhood, and at *Naples*; yet no ſuch as theſe ever ſhook the *Alpes*, or extended to *France*, or other countries remote from *Veſuvius*. The earthquakes produced from the action of volcano's, are confined to a very ſmall ſpace, being properly the effect of the reaction of fire, whereby they ſhake the earth, juſt as a powder magazine when blown up, occaſions a ſhock and a tremor which are felt at many leagues diſtance.

There is yet another kind of earthquakes, very different as to their effects, and probably their cauſes too; ſuch are thoſe which are felt to vaſt diſtances, and ſhake a long ſtretch of ground without the intermediation of any new volcano or eruption. We have examples of earthquakes which were felt at the ſame time in *England*, *France*, *Germany* and *Hungary*; and ſuch are extended greatly more in length than in breadth, and ſhock a belt or zone of earth with a greater and leſs degree of violence in different places, and are almoſt ever accompanied with a dull noiſe like that of a very heavy carriage wheeling on with great rapidity.

To apprehend rightly what are the cauſes of ſuch earthquakes, it muſt be remembered that all ſubſtances which are inflammable and capable of exploſion, do, like powder, at the inſtant of their inflammation, generate a great quantity of air: that air thus generated by fire, is in a ſtate of exceeding great rarefaction, and from its circumſtance of compreſſion within the bowels of the earth, muſt produce moſt violent effects. Suppoſe now that at a conſiderable depth, as a 100 or 200 fathoms, there ſhould happen to be pyrites and other ſulphureous matters, and that through the fermentation excited by the filtration of waters, or by any other means, they come to ignite, let us ſee what will be the conſequence. In the firſt place theſe matters are not diſpoſed regularly in horizontal ſtrata, as ſuch ſubſtances are which have ſettled from

from the ſediment of waters; on the contrary they are in perpendicular fiſſures, in caverns at the foot of ſuch fiſſures, and in other places into which waters can penetrate and there act. Theſe matters taking flame, will produce a great quantity of air, whoſe ſpring compreſs'd in a ſmall ſpace, as that of a cavern, will not only ſhake the ground about it, but will attempt all ways of eſcaping and being at liberty. The paſſages which offer, are the cavities and trenches formed by ſubterraneous waters and rivulets; the rarefied air will be precipitated with violence into every paſſage that is open to it, and form a furious wind, the noiſe whereof will be heard on the earth's ſurface, accompanied with ſhocks and tremors. This ſubterraneous wind generated from fire, will extend full as far as the ſubterraneous caverns or paſſages reach, and excite a tremor, more or leſs violent as it is diſtant from the focus of the conflagration, and meets with paſſages more or leſs confined. This motion being propagated lengthwiſe, the tremor will be ſo too, and will be felt along the extent of a terreſtrial zone; but the air will not be able to produce any eruption or volcano, having found ſpace ſufficient to dilate itſelf in, or becauſe it may have met with ſome vents to eſcape by in the form of wind or vapour: now ſhould it even be denied that any ſubterraneous paſſages do exiſt, through which ſuch wind and vapour can be conveyed, it may notwithſtanding be eaſily conceived that in the very place where the firſt exploſion is made, the ground being elevated to a conſiderable height, it is neceſſary that whatſoever borders upon this

this place muſt be rent, and divided horizontally, and accompany the motion of the firſt blaſt, which will be ſufficient to procure paſſages for communicating the motion to a very great diſtance. This explanation is agreeable to all the phænomena. It is not at the ſame inſtant, nor at the ſame hour that an earthquake is felt in places a 100 or 200 leagues, for example, aſunder: there is neither fire nor eruption above from earthquakes extended to ſo great lengths, and the noiſe which almoſt always accompanies them, marks out the progreſſive motion of the ſubterraneous wind. What has been advanced may be further confirmed by connecting it with other facts; it is known that mines exhale vapours, independently of the winds produced by the current of waters, blaſts of unwholſome and ſuffocating vapour are frequently met with; it is likewiſe well known that there are apertures, abyſſes, and deep lakes which let forth winds at the ſurface, as the lake of *Boleſlaw* in *Bohemia*.

All this being rightly comprehended, I cannot readily diſcern, how it ſhould be believed that earthquakes can produce mountains, ſince the very cauſes of earthquakes themſelves are mineral and ſulphureous matters which are ordinarily found no where but in perpendicular fiſſures and veins of mountains, and other cavities of the earth, moſt of which have been produced by waters; that their ſubſtances by inflaming, produce but a momentary exploſion, and violent winds which follow the tracks of the ſubterraneous waters; that the duration of earthquakes is, in reality, but momentary on the ſurface of the earth, and that conſequently their

their cauſe is no other than an exploſion, and not a durable conflagration; and laſtly, that thoſe earthquakes which ſhake a large ſpace, and extend to mighty diſtances, are ſo far from raiſing ridges of mountains, that they do not ſenſibly elevate the ſurface of the earth, nor form the ſmalleſt hill in the whole length of their courſe.

Earthquakes indeed are by far more frequent in places where there are volcano's, than elſewhere, as in *Sicily* and near *Naples*; 'tis known from obſervations made at different times, that the moſt violent earthquakes happen at the time of the eruption of volcano's; but thoſe earthquakes are not ſuch as extend far, nor can they ever produce a chain of mountains.

It has been ſometimes obſerved that the matters ejected out of *Ætna*, after lying cool for ſeveral years, and being then moiſtened by rains, have rekindled, and thrown out flames with an exploſion ſo violent, as even to produce a kind of little earthquake.

In 1669, during a furious eruption of *Ætna*, which began the 11th of *March*, the ſummit of the mountain ſunk conſiderably, as every one perceived who had ſeen it before [a], which is a proof that the fire of the volcano's proceeds rather from the ſummit than from the interior bottom of the mountain. *Borelli* is of the ſame opinion [b], and ſays expreſsly, that "The fire of volcano's comes "not from the foot nor the center of the moun-

[a] *Philoſ. Tranſ.* abridg'd, vol. ii. p. 387. [b] *De Incendiis Montis Ætnæ.*

"tain,

" tain, but on the contrary from the ſummit, and
" kindles but at a ſmall depth."

Mount *Veſuvius* in its eruptions has often ejected a quantity of boiling water. Mr. *Ray*, who is of opinion that the fire of volcano's comes from a very great depth, ſays that it is ſea water which inſinuates into the internal caverns of the foot of the mountain, and urges for proof the remarkable dryneſs of the ſummit of *Veſuvius*, together with the motion of the ſea, which in violent eruptions recedes from the ſhore; and ſhrinks to that degree, as ſometimes to have left the port of *Naples* in a manner dry: but ſhould theſe facts be true, they would be no ſolid proof that the fire of volcano's comes from a very great depth: for the water they throw out is certainly rain water which ſoaks in through fiſſures, and is collected in the cavities of the mountain: freſh ſprings and brooks are ſeen to run from the ſummits of volcano's, in the ſame manner as from other high mountains; and as they are hollow, and have undergone more concuſſions than other mountains, it is not ſtrange that waters ſhould be depoſited in the caverns within them, and that thoſe waters ſhould be rejected, with other ſubſtances, during their eruptions. As to the motion of the ſea, it ariſes ſolely from the ſhock communicated to its water by the exploſion, which muſt occaſion an afflux and reflux, according to different circumſtances.

The ſubſtances which volcano's reject, iſſue out moſt commonly under the form of a torrent of melted minerals, which inundates all places round ſuch mountains: thoſe rivers of liquified matter ſtretch

ſtretch to conſiderable diſtances, and in cooling, form themſelves into horizontal or inclining beds, which as to their poſition are ſimilar to the beds which are made of the ſediments of waters; but it is very eaſy to diſtinguiſh the beds formed by the ſpreading of ſubſtances rejected by volcano's, from thoſe which ariſe from ſediments of the ſea. 1*ſt*, Becauſe they are not every where of an equal thickneſs. 2*d*, Becauſe they contain no other than ſuch matters as may be evidently perceived to have been calcined, vitrified, or melted. 3*d*, Becauſe they do not extend to a very great diſtance. There being a multitude of volcano's in *Peru*, and the foot of moſt of the volcano's of the *Cordilieres* covered with matters vomited out of thoſe mountains, it is not ſurprizing that no ſea ſhells ſhould be found in all that ſoil, ſince they have been calcined and deſtroyed by the action of the fire: but I am perſuaded, that were one to dig into the clayey ſoil, which according to M. *Bouguer*, is the ordinary land of the valley of *Quito*, ſhells would be met with there, as they are in all other places; ſuppoſing that ſoil to be really of clay, and that it is not formed, as is that at the foot of the mountains, of the excrements of volcano's.

It has been often aſked, for what reaſon are volcano's found in high mountains? I think I have in part ſatisfy'd this query already elſewhere, however I will not cloſe this ſubject without explaining myſelf more particularly.

The pikes or points of mountains were all of them once covered and invironed with ſand and earth,

earth, which rain waters afterwards washed down into the valleys, and left nothing but the rocks or stones remaining, which formed the kernel or core of the mountain; this core being laid bare, and stripped to the foot, became after this liable to further injuries from the air, besides the scaling off and separating of many great and small fragments by frost, which rowled down below, several rocks of the summit cleaving asunder from the same cause. Those which formed the base of the summit being uncovered, and no more supported by the surrounding earth, gave way a little, and by separating from each other formed small interstices: this yielding of the lower rocks could not take place without communicating a more considerable motion to the upper ones, whereby they were cleft and rent from one another. In consequence of all this an infinity of perpendicular fissures great and small, came to be wrought in the core of the mountain, from the summit to the base of the lower rocks: through these the rains penetrated, and loosened or dissolved all the minerals and other substances in the heart of the mountain, which they were capable of acting upon; they formed pyrites, sulphurs and other combustible matters; and when in process of time these matters became accumulated in a large quantity, they fermented, took flame and produced explosions and other effects of volcano's. Perhaps too there might be a stock of such mineral substances already formed in the heart of the mountain even before the rains had penetrated, and these might force open fissures, and give passages to the water and air, which put them into the ſtate of inflammation which produced a volcano. No ſuch motions can be brought about in plains, where all things ſubſiſt in a perfect repoſe, and nothing is capable of being diſplaced, ſo that it is not at all ſtrange that they are entirely free from volcano's.

When coal mines are opened, which are uſually found in clayey ſoils at a great depth, it ſometimes happens that the beforementioned ſubſtances take fire, and there are ſome mines in *Scotland*, *Flanders*, &c. which continue burning ſeveral years: the communication of the air is ſufficient for this effect: but this ſort of fire produces but ſlight exploſions, without forming volcano's, becauſe all being ſolid and compact in ſuch places, no ſuch fires can be wrought up as thoſe are in volcano's, where there are cavities into which the air enters, and by augmenting and aſſiſting the action of the fire, produces the terrible effects we have treated of in this eſſay.

OF

A SUMMARY of the CAUSES OF THE ALTERATIONS Which have happened to the Face of the EARTH.

THE changes and alterations that have been made in the ſuperficial part of the terraqueous globe have been effected chiefly by water, fire and wind. Thoſe by water have been either by the motions of the ſea, or by rains; and both either ordinary or extraordinary: the ordinary tides and ſpring-tides of the ſea do waſh away the ſhores, and change ſand-banks and the like. The extraordinary and tempeſtuous motions of the ſea, raiſed by raging and impetuous winds, ſubterraneous fires, or ſome other hidden cauſes, overwhelm iſlands, open fretum's, throw up huge beds and banks of ſand, nay vaſt baiches (*beaches*) of ſtone, extending ſome miles, and drown whole countries. The ordinary rains contribute ſomething

thing to the daily diminution of the mountains, filling up of the valleys, and atterating (*wearing away*) the ſkirts of the ſeas. The extraordinary rains cauſing great floods and deluges, have more viſible and remarkable influences upon ſuch mutations, doing that in a few days, which the ordinary weather could not effect, it may be, in an hundred years.

In all theſe changes the winds have a great intereſt; the motion of the clouds being wholly owing to them, and in a great meaſure alſo the overflowings and inundations of the ſea.

Whatever changes have been wrought by earthquakes, thunders, and eruptions of volcano's, are the effects of fire.

All theſe cauſes co-operate towards the lowering of mountains, leveling of the earth, ſtraitening and landing up of the ſea, and in fine compelling the waters to return upon the dry land, and cover the whole ſurface of it, as at the firſt. How to obviate this in a natural way, I know not, unleſs by a tranſmutation of the two elements of water and earth one into another, which I can by no means grant. 'Tis true indeed, the rocky parts of the mountains may be ſo hard and impenetrable, as to reſiſt and hold out againſt all the aſſaults of the water, and utmoſt rage of the ſea; but then all the earth and ſand being waſhed from them, nothing, but as it were their ſkeletons, will remain extant above the waters, and the earth being in effect drowned.

But though I cannot imagine or think upon any natural means to prevent and put a ſtop to this effect, yet I do not deny that there may be ſome; and I am the rather inclinable ſo to think, becauſe the world doth not in any degree proceed ſo faſt towards this period, as the force and agency of all theſe cauſes together ſeem to require. For, as I ſaid before, the oracle predicting the carrying on the ſhore of *Cilicia*, as far as *Cyprus* by the earth and mud that the turbid river *Pyramus* ſhould bring down, and let fall in the interjacent ſtrait, is ſo far from being filled up, that there hath not any conſiderable progreſs been made towards it, ſo far as I have heard or read, in theſe 2000 years. And we find by experience, that the longer the world laſts, the fewer concuſſions and mutations are made in the upper or ſuperficial region of the earth; the parts thereof ſeeming to tend to a greater quiet and ſettlement.

Beſides the ſuperficies of the ſea, notwithſtanding the overwhelming and ſubmerſion of iſlands, and the ſtraitening of it about the outlets of rivers; and the earth it waſhes from the ſhores ſubſiding, and elevating the bottom, ſeems not to be raiſed higher, nor ſpread further, or bear any greater proportion to that of the land than it did a 1000 years ago.

SOME

SOME CONSIDERATIONS
ON THE
Causes of EARTHQUAKES.

IN the earthquake which happened the 8th of *March* 1749-50, I being then awake in bed, on a ground-floor, near the church of *St. Martin's in the Fields*, very sensibly felt the bed heave, and consequently the earth must heave too. There was a hollow, obscure, rushing noise in the house, which ended in a loud explosion up in the air, like that of a small cannon: the whole duration, from the beginning to the end of the earthquake, seemed to be about four seconds of time. The soldiers who were upon duty in *St. James's Park*, and others who were then up, saw a blackish cloud, with considerable lightning, just before the earthquake began; it was also very calm weather.

In the history of earthquakes it is observed, that they generally begin in calm weather with a black cloud. And when the air is clear, just before an earthquake, yet there are often signs of plenty of inflammable sulphureous matter in the air; such as *ignes fatui* or *jack-a-lanterns*, and the meteors called falling stars.

Now, I have shewn many years since, in the appendix to my *Statical Essays*, experiment 3, p. 280. the effect that the mixture of a pure and a sulphureous air have on each other; *viz.* by turning the mouth downwards into a pan of water, of a glafs veffel of a capacity fufficient to hold about two quarts, with a neck about 20 inches long, and two inches wide. Then, by putting under it, in a proper glafs veffel, with a long narrow neck, a mixture of *aqua fortis*, and powdered *pyrites*, *viz*. the ftone of which vitriol is made, there will be a brifk ferment, which will fill the glafs with reddifh fulphureous fumes; which by generating more air than they deftroy, will caufe the water, with which the whole neck of the glafs veffel was filled, to fubfide confiderably. When the reddifh fulphureous air in the upper part of the glafs is clear, by ftanding two or three hours, if then the mouth of the inverted glafs is lifted out of the water, fo as to let the water in the neck of the glafs fall out; which, fupoofing it to be a pint, then an equal quantity of frefh air will rufh in at the mouth of the neck of the veffel, which muft be immediately immerfed in the water: and upon the mixture of the frefh air with the then clear fulphureous air, there will inftantly arife a violent agitation between the two airs, and they will become, from tranfparent and clear, a reddifh turbid fume, of the colour of thofe vapours, which were feen feveral evenings before the late earthquake, during which effervefcence, a quantity of air, nearly equal to what frefh air was let in, will be deftroyed; which is evident by the rifing up of the water in the neck of the glafs, almoft as high as before. And if, after the effervefcence of the mixed airs is over, and they become clear again, frefh air be admitted, as before; they will again grow reddiſh and turbid, and deſtroy the new admitted air, as before; and this after ſeveral repeated admiſſions of freſh air: but after every readmiſſion of freſh air, the quantity deſtroyed will be leſs and leſs, till no more will be deſtroyed. And it is the ſame after ſtanding ſeveral weeks, provided in the mean time, too much freſh air had not been admitted. Now, I found the ſum total of the freſh air thus deſtroyed to be nearly equal to the firſt quantity of ſulphureous air in the inverted glaſs.

Since we have in this experiment a full proof of the briſk agitation and efferveſcence which ariſes from the mixture of freſh air with air that is impregnated with ſulphureous vapours, which ariſe from ſeveral mineral ſubſtances, eſpecially from the *pyrites*, which abounds in many parts of the earth; may we not with good reaſon conclude, that the irkſome heat, which we feel in what is called a cloſe ſultry temperature of the air, is occaſioned by the inteſtine motion between the air and the ſulphureous vapours which are exhaled from the earth? which efferveſcence ceaſes as ſoon as the vapours are equably and uniformly mixed in the air; as happens alſo in the efferveſcences and fermentations of other liquors. The common obſervation therefore, that lightning cools the air, ſeems to be grounded on good reaſon; that being the utmoſt and laſt effort of this efferveſcence.

May we not hence alſo, with good probability, conclude, that the firſt kindling of lightning is effected by the ſudden mixture of the pure ſerene air above the clouds, with the ſulphureous vapours, which are ſometimes raiſed in plenty, immediately below the clouds? the moſt dreadful thunders being uſually when the air is very black with clouds; it rarely thundering without clouds; clouds ſerving, in this caſe, like the abovementioned inverted glaſſes, as a partition between the pure and ſulphureous airs: which muſt therefore, upon their ſudden admixture through the interſtices of the clouds, make (like the two airs in the glaſs) a more violent efferveſcence, than if thoſe airs had, without the intervention of the clouds, more gradually intermixed, by the conſtant more gradual aſcent of the warmer ſulphureous vapours from the earth, and the deſcent of the cold ſerene air from above. And though there was no luminous flaſh of light in the glaſs, yet, where ſuch ſudden efferveſcence ariſes, among a vaſt quantity of ſuch vapours in the open expanſe of air, it may, not improbably, acquire ſo rapid a velocity, as to kindle the ſulphureous vapours, and thereby become luminous.

And ſince, from the effects that lightning is obſerved to have on the lungs of animals, which it often kills, by deſtroying the air's elaſticity in them, as alſo from its burſting windows outwards, by deſtroying the air's elaſticity on the outſide of thoſe windows: ſince, I ſay, it is hence probable, that the ſulphureous fumes do deſtroy a great quantity of elaſtic air; it ſhould therefore cauſe great commotions and concuſſions in the air, when the air ruſhes into thoſe evacuated places; which it muſt naturally do with great velocity.

Dr. *Papin* has calculated the velocity with which air

air rushes into an exhausted receiver, when driven by the whole pressure of the atmosphere, to be at the rate of 1305 feet in a second of time; which is at the rate of 889 feet in an hour: near 18 times a greater velocity than that of the strongest storms, which is estimated to be at the rate of 50 miles in an hour [a]

Hence we see that an outrageous hurricane may be caused, by destroying a small portion of the elasticity of the air of any place, in respect to the whole. No wonder then, that such violent commotions of the air should produce hurricanes and thunder-showers; especially in the warmer climates; where both the sulphureous and watery vapours, being raised much higher, and in greater plenty, cause more violent effects.

Monsieur *de Buffon*, in his *Natural history and theory of the earth*, mentions black dark clouds in the air, near the tempestuous *Cape of Good Hope*, and also in the ocean of *Guiney*, called by the sailors the *Ox's Eye*, which are forerunners of terrible storms and hurricanes. Whence it is to be suspected, that they are large collections of sulphureous vapours; which, by destroying suddenly a great quantity of the elastic air, cause the ambient air to rush with great violence into that vacuity, thereby producing tempests and hurricanes; and off the coast of *Guiney* they have sometimes three or four of these hurricanes in a day; the forerunners of which are these black sulphureous clouds, with a serene clear air and calm sea; which on a sudden turns tempestuous, on the explosion of

[a] *Phil. Trans. n.* 184, *p.* 195.

theſe ſulphureous clouds. And in *Jamaica* they never have an earthquake when there is a wind to diſperſe the ſulphureous vapours.

In the like manner we find, in the late earthquakes at *London*, and in the accounts of many other earthquakes, that before they happen, there is uſually a calm air, with a black ſulphureous cloud: which cloud would probably be diſperſed like a fog, were there a wind: which diſperſion would prevent the earthquake, as it is probably cauſed by the exploſive lightning of this ſulphureous cloud; being both nearer the earth, than common lightnings, and alſo at a time when ſulphureous vapours are riſing from the earth in greater quantities than uſual, which is often occaſioned by a long ſeries of hot and dry weather. In which combined circumſtances, the aſcending ſulphureous vapours in the earth may probably take fire, and thereby cauſe an earth-lightning; which is at firſt kindled at the ſurface, and not at great depths, as has been thought: and the exploſion of this lightning is the immediate cauſe of an earthquake.

It is in the like manner that thoſe meteors, which are called falling ſtars are ſuppoſed to be kindled into a flame at the upper part of a ſulphureous train, which is kindled downwards into a flame, in the ſame manner as a freſh-blown-out candle is inſtantly lighted from another candle held over it at a diſtance, in the ſulphureous inflammable ſmoke of it.

I am ſenſible that it may ſeem improbable, that the aſcending ſulphureous vapours in the earth ſhould thus be kindled; but, ſince they are con-

tn ually

tinually aſcending through the pores of the earth, more or leſs, for many good and uſeful purpoſes, it is plain there is room for them to paſs. Beſides, as Monſ. *de Buffon* remarks, naturaliſts have obſerved perpendicular and oblique clefts, in all kinds of layers of earth, not only among rocks, but alſo among all kinds of earth, that have not been removed, as is obſervable wherever the earth is open to any depth. Now theſe clefts are cauſed by the drying of the ſeveral horizontal layers of the earth; and will alſo be conſiderably the wider in long, dry, hot ſeaſons, which are uſually the preparatory forerunners of earthquakes, and the exploſion of the ſulphureous vapours may probably widen them the more.

It is very obſervable, in the opinion of *Borelli*, and other naturaliſts, that volcano's begin firſt to kindle near the ſurface or top of the mountains, and not in the caverns in the lower parts of the mountains. Monſ. *de Buffon* ſays that earthquakes are moſt frequent where there are volcano's, ſulphureous matter abounding moſt there: but that, though they continue burning long, yet they are not very extenſive: but that the other ſort of earthquakes, which are not cauſed by a volcano, extend often to a great diſtance. Theſe are much longer eaſt and weſt, than broad north and ſouth; and ſhake a zone of earth with different degrees of force in different parts of their courſe: *viz.* in proportion to the different quantities of exploſive ſulphureous matter in different places. Theſe kind of earthquakes are obſerved to be progreſſive, and to take time to extend to the great diſtances,

tances, ſometimes of ſome thouſands of miles. They are an inſtantaneous exploſion in every place, near the ſurface of the earth; and therefore do not produce mountains, and iſlands, as volcano's ſometimes do.

The earthquake in *London*, *March* 8, 1749-50, was thought to move from eaſtward to weſtward. M. *de Buffon* mentions an earthquake at *Smyrna* in the year 1688, which moved from weſt to eaſt; *viz.* becauſe the firſt kindling probably began on the weſtern ſide; and in the earthquake at *London* on the eaſtern ſide. And accordingly it was obſerved that the reddiſh bows in the air, which appeared ſeveral days before that earthquake, aroſe in the eaſt, and proceeded weſtward. It was obſerved after the earthquake at *Smyrna*, that the caſtle walls which run from eaſt to weſt, were thrown down, but thoſe from north to ſouth ſtood; and that the houſes on rocks ſtood better than thoſe on the earth.

M. *de Buffon* relates, that the vibrations of the earth in earthquakes, have commonly been from north to ſouth, as appears by the motion of the lamps in churches: which makes it probable, that tho' the progreſs of the earthquake at *Smyrna* was from weſt to eaſt, yet the vibrations of the earth might be from north to ſouth, and thereby occaſion the fall of the caſtle walls which ran from eaſt to weſt, but not thoſe which ran from north to ſouth: A probable argument, that as the freeſt paſſage, ſo the greateſt exploſions were made in the clefts of the earth which ran eaſt and weſt; which would make the vibrations north and ſouth.

It

It was obſerved that the waters turned foul the day before an earthquake at *Bologna*, in *Italy*: and I was informed, that the water of ſome wells in *London* turned foul at the time of the earthquakes; which was probably occaſioned by the aſcent of great plenty of ſulphureous vapours thro' the earth.

As to the hollow rumbling noiſe which is uſually heard in earthquakes, it ſeems not improbable that it may be occaſioned by the great agitation that the electrical æthereal fluid is put into by ſo great a ſhock of a large maſs of earth. For if the like motion of a ſmall revolving glaſs globe can excite it to the velocity of lightning, and that with a force ſufficient to kill animals; how much greater agitation may it probably be excited to, by the exploſive force of an earthquake!

The exploſion of cannon in St. *James's Park* is obſerved to electrify the glaſs windows of the *Treaſury*. And what makes it ſtill more probable, is, the analogy that there is between them in other reſpects. For as the electrical flaſh ruſhes with the velocity of lightning, along the moſt ſolid bodies, as iron, &c. and as I have ſeen it run only on the irregular gilding of leather; ſo ſuch ſolid bodies are obſerved to be the conductors of aereal lightning, which rends oaks in pieces, and has been known to run along and melt an iron bell-wire on two ſides of a room, &c. And accordingly it was obſerved, in the great earthquake at *Jamaica*, that the moſt tremendous roaring was in the rocky mountains. And in the late earthquake

I of

of *March* 8 in *London*, the loudeft explofions were thought to be heard near fuch large ftone buildings as churches, with lofty fteeples and fpires.

I, who lay in *Duke's Court* near St. *Martin*'s church, and was awake all the time of the earthquake, plainly heard a loud explofion up in the air, like that of a fmall cannon: which made me conjecture, that the noife was owing to the rufhing off, and fudden explofion of the electrical fluid, at the top of St. *Martin*'s fpire; where all the electrical effluvia, which afcended up along the larger body of the tower, being by attraction ftrongly condenfed, and accelerated at the point of the weather-cock, as they rufhed off, made fo much the louder expanfive explofion.

THE

THE PHILOSOPHY OF EARTHQUAKES.

Positions or Circumstances.

I. THAT earthquakes always happen in calm, warm, dry, sultry seasons; or in a dry frosty air.

II. That they are felt at sea, or on lakes, rivers, even in the main ocean, as well as on land: and at that time the sea and waters are calm.

III. That earthquakes differ very much in magnitude. Some shake a very large tract of country, at the same instant of time: sometimes extend to many countries, separated by mountains, lakes, seas, the ocean.

IV. That earthquakes differ much in the quantity of their vibratory motion; whence in some, though largely extended, they are innocuous: in others, both small and large, they lay all in ruins.

V. That a hollow thundering noise accompanies them, or rather seems to precede the shock; which rolls in the air, like the noise of cannon.

VI. That they are felt more sensibly in the upper story of houses, than in the lower. On lofty

lofty buildings, ſteeples, *Turkiſh* minorets, and the like.

VII. That the ſhock is more violent upon more ſolid buildings, churches, caſtles, towers, and ſtone houſes, than on thoſe of ſlighter materials.

VIII. That many people find themſelves ſick at ſtomach, with head-achs, *vertigo's*, pains in their joints, and the like: which ſometimes laſt for the day after, or longer.

IX. That earthquakes generally happen to great towns and cities: and more particularly to thoſe that are ſituated on the ſea, bays, and great rivers.

X. That earthquakes do not cauſe any damage to ſprings, and fountains: but the water in wells becomes foul for ſome time.

XI. That they are frequent in the neighbourhood of a volcano.

XII. That earthquakes often ſhake rocks, mountains, cliffs hanging over the ſea, ſplit them from top to bottom, throw down great parcels of them.

XIII. That fowls domeſtic, birds in the air, cattle in the fields are affrighted, fiſhes in the water much affected therewith.

XIV. That chandeliers in churches vibrate, bells in ſteeples and houſes ring.

XV. That ſometimes the hollow, thundering noiſe accompanying an earthquake, is heard without any motion of the earth: at other times accompanies it.

XVI. That fire balls and meteors are frequently obſerved then.

XVII.

XVII. That the ſurface of the earth is chiefly, and moſt frequently the object of earthquakes.

XVIII. That earthquakes affect to run up rivers and ſea-ſhores, and act more violently on places neighbouring thereto.

As to the cauſe of earthquakes, the moderns have not improved upon the ancients, any farther than by the fancied analogy of ſome chymical experiments. But theſe chymical experiments, and all ſorts of exploſions by gun-powder and the like, are to me a very unſatisfactory ſolution. They are merely artificial compoſitions, which can have nothing ſimilar, in the bowels of the earth, and they produce their effects by violence, by rending and tearing, by a *ſolutio continui*. This indeed is too often the caſe of earthquakes, but that in a partial degree, not at all equivalent to the compaſs of the ſhock; and is very far from being the conſtant concomitant of an earthquake; quite the contrary. Innumerable ſuch happen where there is no breach of the ſurface; and in the three or four felt by us of late years, nothing of it has appeared. But the immenſity of the vibration of the earth which ſhook every houſe in *London*, with impunity, and for twenty miles round, can never, in my apprehenſion, be owing to ſo unbridled a cauſe, as any ſubterraneous vapours, fermentations, rarefactions, and the like; the vulgar ſolution. Nor does the kind of motion which I diſcern in an earthquake, in any ſort agree with what we ſhould expect from exploſions.

The

The ſtruggles of ſubterraneous winds and fires, that ſhould heave up the ground, like animal convulſions, ſeem to me impoſſible: Their powers, and manner of acting, if ſuch there be, are quite incapable of producing the appearance of an earthquake. That theſe ſhould operate inſtantaneouſly, in one minute through a circle of 30 or 40 miles diameter, or more, I could not conceive: nor that there ſhould be any poſſible, much leſs ready paſſage through the ſolid earth, for ſuch nimble agents, as every one is apt to imagine, that ſpeak of this appearance; without ſufficiently reflecting on the innumerable difficulties in that *hypotheſis*.

We cannot pretend to deny, that there may be ſuch vapours, and fermentations, inflammable ſubſtances, and actual fires in the bowels of the earth, and that there may be ſome caverns underground, as well as we find ſome few above ground: ſuch as *Pool's Hole*, the *D—l's A—ſe* in the *Peak* of *Derbyſhire*, and *Okey-Hole* in *Somerſetſhire*. Theſe, I believe, to have been ſo from the creation, and never were made by earthquakes. We know there are hot ſprings running continually: there are ſome volcano's frequently belching out flames and ſmoke, and to theſe perhaps ſome earthquakes may be owing, though not according to the vulgar notion.

But theſe matters are very rare, and much rarer than earthquakes, both as to time and place. *Veſuvius* in *Italy*, and in that part of it abounding with mines of ſulphur: *Ætna* in *Sicily*, and *Hecla* in *Iceland*; theſe are all we know of in the old world. In the *Andes* mountains of *America* there

there are ſome. The ſcarcity of theſe appears to me a ſtrong argument againſt the common deductions made therefrom, as to their being the cauſe of earthquakes. And further, we cannot poſſibly think of earthquakes doing their work that way, without abſolutely ruining the whole ſyſtem of ſprings and fountains, throughout the whole country where they paſs. But all this is quite contrary to fact.

Theſe conſiderations I apply only to this little inconſiderable ſpace of a circle of 30 miles diameter, as with us. But what is that to the earthquakes we read of in hiſtory? In the year of our Lord 17, no leſs than thirteen great and noble cities in *Aſia Minor* were deſtroyed in one night. The compaſs of this earthquake may be reckoned to take up 300 miles diameter, as a circle. And altogether as great, nay far greater in extent was that moſt dreadful one of *November* 1, 1755, whereby, as of old the cities of *Aſia*, *Liſbon* was deſtroyed, with ſeveral in *Africa*, and a vaſt number beſides nigh totally ruined: yet none of theſe were ſwallowed up, but ſhaken into an heap of rubbiſh.

From theſe conſiderations I cannot perſuade myſelf to entet into the opinion of vapours and eruptions being the cauſe ſought for. If we would conſider things like philoſophers, let us propoſe to ourſelves this problem: *Where is the power to be placed, that is required to move a ſurface of Earth, thirty miles in diameter?*

To anſwer this, conſult the ingineers, and thoſe that make mines in the ſieges of towns; they will acquaint us, that the effect of mines is produced in form of an inverted cone; and that a diameter of 30 miles, in the base, will require an axis of 15 or 20 miles to operate upon that base, so as to shake it a least. Now the vapours, or whatever power we propose to operate, according to the foregoing requisite, in order to form the appearance of an earthquake, must be 15 or 20 miles deep in the earth. But what mind can conceive, that any natural power is able to move an inverted cone of solid earth, whose base is 30 miles diameter, whose axis is 20? or, was it possible, would not the whole texture of that body of earth be quite disturbed and shattered, especially in regard to its springs and fouutains? but nothing like this is ever found to be the consequence of an earthquake, though fatal to cities.

Apply this reasoning to the earthquake of *Asia minor*; and this vigorous principle at the apex of the cone must lie, at least, 200 miles deep in the ground: enough to shew the absurdity of any moving power placed under the earth! a cone of 300 miles diameter at base, 200 in axis. I dare be bold to say, that all the gun-powder made since its invention, if put together and fired, would not be able to move it: how much less pent up vapours? what must we say of a circle 900 miles diameter?

But could that be admitted as possible; would any one be persuaded, that such a subterraneous tumult, of so vast an extent, will be no ways injurious to the internal system of springs and fountains, and that this shall be often repeated without

out the leaſt damage? we may as well imagine, that we can ſtab a man a hundred times, and never touch vain or artery.

We are then next to inquire: *What is the cauſe of earthquakes?*

In an age when *electricity* has been ſo much our entertainment, and our amazement; when we are become ſo well acquainted with its ſtupendous powers and properties, its velocity and inſtantaneous operation through any given diſtance; when we ſee, upon a touch, or an approach, between a non-electric and an electric body, what a wonderful vibration is produced! what a ſnap it gives! how an innocuous flame breaks forth! how violent a ſhock! is it to be wondered at, that hither we turn our thoughts, for the ſolution of the prodigious appearance of an earthquake?

Here is at once an aſſemblage of all thoſe properties and circumſtances which we ſo often ſee in courſes of electricity. Electricity may be called a ſort of ſoul to matter; thought to be an ethereal fire pervading all things; and acting inſtantaneouſly, where, and as far as it is excited.

We had lately read at the royal ſociety a very curious diſcourſe from Mr. *Franklin* of *Philadelphia*, concerning thunder-guſts, lightning, the northern lights, and like meteors; all which he rightly ſolves from the doctrine of electricity. For, if a cloud raiſed from the ſea, which is a non-electric, happens to touch a cloud raiſed from exhalations of the land, when electriſed, it muſt immediately cauſe thunder and lightning. The electrical fire flowing from the touch of perhaps a thousand miles compass of clouds, makes that appearance which we call lightning. The snap which we hear in our electrical experiments, when reechoed from cloud to cloud, the extent of the firmament, makes that affrightning sound of thunder.

From the same principle I infer, that, if a non-electric cloud discharges its contents upon any part of the earth, when in a high electrified state, an earthquake must necessarily ensue. The snap made upon the contact of many miles compass of solid earth, is that horrible uncouth noise, which we hear upon an earthquake; and the shock is the earthquake itself.

In the relation from *Portsmouth*, and the *Isle of Wight*, concerning the shock of the earthquake on the 18th of *March*, 1749-50, the writer observes, the day was warm and serene; but upon a gentle shower falling in the evening, the earthquake came. Here we have reason to apprehend the electrified state of the earth, and the touch of the non-electric, which caused the earthquake,

The learned Dr. *Childrey* observes, treating on this subject, that earthquakes happen upon rain, in the time of a great drought.

'Tis objected, that, if this were the case, nothing would be more frequent than earthquakes: but these two circumstances concurring, a shower and dry weather must not necessarily cause it, any more than touching a tube before it is electrified causes a snap. The earth must be in a proper electrified state to produce it; and electricity has its fits; is remitted, intended, ceased and re-

recommenced. It has its bounds. All causes must concur: though a shower of rain falling upon the earth, when electrified, may cause an earthquake, yet too much rain before, will prevent that state of electricity necessary.

The day before the catastrophe of *Port Royal*, the weather was remarkably serene and clear. In that most dreadful earthquake of *Sicily*, 1692, where 54 cities and towns, besides a great number of villages, were destroyed, but especially the whole city of *Catanea*; it was preceded by a most agreeable, serene, and warm season, which was the more observable, on account of its being unusual at that time of the year.

I have been informed, that in the mornings of our earthquakes in *February* and *March* 1749-50, the air was serene and calm, and on the morning before that, in *February*, the air was observed to be perfectly calm; and that a little before, a black cloud appeared over great part of the horizon. Dr. *Hales* says the centinels in St. *James*'s park, and others who were abroad in the morning of *March* the 8th, observed a large black cloud, and some corruscations, just before the shock; and that it was very calm weather: and that in the history of earthquakes, they generally begin in calm weather, with a black cloud.

We have been acquainted by those who remember it, that in the earthquake of *November* 1703, which happened in *Lincolnshire*, the weather was calm, close, gloomy, warm, and dry; in a degree highly unusual at that season. And thus was it with us all the year 1749, thereby preparing the earth's ſurface for the electrical ſtroke, which I have aſſerted to be the cauſe of them.

In the account of the great earthquake of *November* 1, 1755, from *Amſterdam*, it was wrote that the weather was calm ; the like from *Berlin*, *Kinſale*, *Gibraltar*, *Liſbon*, &c.

Mr. *Flamſteed* ſuppoſes a calm even neceſſary before an earthquake : And Dr. *Hales* ſays, that long, dry, hot ſeaſons are uſually the preparatory fore-runners of earthquakes.

This obſervation precludes the ſuſpicion of earthquakes ariſing from tumults and commotions in the upper, or under region of the air. The remarkable clearneſs of the air before earthquakes, obſerved by all, ſhews evidently how free it is from vapours, or the like.

Agreeable to our fifth poſition, Mr. *Flamſteed* writes [a], " a hollow noiſe in the air always precedes an earthquake, ſo near, that it rather ſeems to accompany it," this he ſpoke of that felt in *London* 1692, when the noiſe was heard by many that lived in the out-ſtreets and alleys, remote from the conſtant tumult of the great ſtreets ; but in both our latter ones, the whole city heard the noiſe.

A gentleman of *Hartingfordbury* ſays, the noiſe preceded the ſhock. And this is a common obſervation, which at once both ſtrengthens our opinion of electricity, and confutes that of ſubterraneous vapours ; for, in the latter caſe, the concuſſion muſt precede the noiſe.

[a] Letter concerning an Earthquake.

Juſt

Juſt before the earthquake of *March* 8, 1749-50 Mr. Secretary *Fox*'s ſhepherd at *Kenſington* was ſurprized with a very extraordinary noiſe in the air, rolling over his head, as of cannon cloſe by. He likewiſe thought that it came from the north-weſt, and went to the ſouth-eaſt; a motion quite contrary to what muſt have been the caſe, if it were really cannon. It paſſed ruſhing by him, and inſtantly he ſaw the ground, a dry and ſolid ſpot, wave under him like the face of a river. The trees of the avenue nodded their tops, and were ſhaken like ſpears.

In the earthquake of *September* 30, 1750, they were ſuddenly ſurprized with an uncommon noiſe in the air, like the rolling of large carriages in the ſtreets, for about 20 ſeconds. At the ſame inſtant they felt a great ſhock or ſnap, which ſenſibly ſhook a punch bowl, and made it ring.

Agreeable to our ſecond poſition, Mr. *Flamſteed* " writes [b], that earthquakes are felt at ſea equally " as at land. Our merchants ſay, that tho' the " water in the bay of *Smyrna* lies level, and " ſmooth as a pond, yet ſhips riding there feel " the ſhocks very ſenſibly; but in a very dif- " ferent manner from the houſes at land: For " they heave not, but tremble; their maſts " ſhiver, as if they would fall to pieces, and their " guns ſtart in their carriages, though the ſurface " of the ſea be all the time calm and unmoved.

Dr. *Hooke* tells us [c], " that a ſhip felt a ſhock " in the main ocean; that the paſſengers, who

[b] Letter concerning an earthquake.
[c] Philoſ Collections, N° 6. p. 185.

" had been asleep in their cabins, came upon deck
" in a fright, fearing the ship had struck upon
" some rock; but on heaving the lead, found
" themselves out of all soundings."

In the earthquake of the first of *November* 1755, the *Dutch* ships of cape St. *Marys* fired guns of distress, thinking they struck on ground; and we have received many accounts of ships at sea, in the middle of the ocean, participating in this amazing phenomenon. A very loud thundering noise begins it; it seems as if casks were rolling about the deck. The masts, the whole ship trembled like a reed shaken with the wind. A great thump felt at the bottom of the ship, as if struck upon a rock. The compass often overturned in the benacle, fire-balls and flashes of lightning seen by night.

All this is extremely agreeable to our assumption. The water receives the electrical touch, and vibratory intestine motion of its parts, as well as land: And the impression may be made solely on the water, a non-electric, by the touch of an electric fire-ball, or the like; and that seems to have been often the case. The proper vibratory motion is unpressed on the water without ruffling its surface; and so communicated to all the parts of the ship, gives the sense of a shock to the bottom, the shivering to the mast, and the rest of the symptoms; which sufficiently proclaim the cause of it to be an electrical impression upon the water. The president of the Royal Society mentioned a relation of a waterman that felt it in his boat upon the river: he thought it like a great thump at the bottom

bottom of the boat. And ſo the ſhips at ſea fancy they ſtrike upon a rock.

This makes us apprehend the reaſon of the fiſhes leaping out of the canal in *Southwark*, of which we had an account. So, in that of *Oxford* 1683, one fiſhing in the *Charwell* felt his boat tremble under him, and the leſſer fiſhes ſeemed affrighted by an unuſual ſkipping. That electricity is the cauſe ſought for, ſeems deducible from this conſideration. Several writers on earthquakes aſſimilate theſe vibrations of the earth to thoſe of a muſical ſtring: experiments have ſhewn, that fiſhes in water may be killed, by the particular tone of a muſical ſtring; and 'tis known that electricity will kill animals. They aſſuredly felt the vibrating motion in the water, which they were abſolutely ſtrangers to before. No doubt it made them ſick; as thoſe of weak nerves on land. And this circumſtance alone precludes any ſuſpicion of ſubterraneous fires under the ocean. Or, if we were to admit of it, would the boiling of the water exhibit any appearance, like what we are ſpeaking of, either to the water, or to the ſhip?

Mr. *Flamſteed* likewiſe concurs in our eighth poſition [d], "that many people found themſelves "ſick at ſtomach, and their heads dizzy and "light; ſo that thoſe that had formerly fits of "apoplexies, dreaded their return; particularly "one gentleman, a ſurgeon, feeling himſelf ſo "affected, and fearing a return of his apoplexy, 'reſolved to be let blood, without ſuſpecting the "earthquake."

[d] Letter concerning an earthquake.

After the two ſhocks we felt in *February* and *March* 1749-50. Many people had pains in their joints and back, as after electrifying; many had ſickneſs and head-achs, hyſteric and nervous diſorders, and cholicks, for the whole day after, and ſome much longer, eſpecially people of weak nerves, weak conſtitutions; ſome women miſcarried upon it; to ſome it has proved fatal.

To this we muſt attribute that relation we had of the dog lying aſleep before the fire; but upon the earthquake, he ſuddenly roſe up, run about the room, whining and endeavouring to get out.

Any ſolid matter is capable of being put into a ſtate of electricity, ſuch as iron guns; and the more ſo, by reaſon of their ſolidity: and in proportion to it, is the greatneſs of the ſnap, and of the ſhock; and a kind of lambent flame iſſues from the point of contact; and likewiſe ſomewhat of a ſulphureous ſmell: ſo that if both flame and ſmell were diſcernible in an earthquake; 'tis to be found without going to the bowels of the earth.

Dr. *Hales* mentions that ſolid bodies are the beſt conductors of aerial lightening; whence oaks are rent, and iron melted. And in our earthquakes in *London*, the loudeſt noiſe was heard near ſuch large ſtone buildings, as churches with lofty ſteeples. From the top of theſe we muſt apprehend, that the electrical exploſion goes off into the open air; as in our experiments, from the points of ſwords, and the like.

The electrical ſhock is proportionate to the ſolid electrified, agreeable to our ſeventh poſition. This

This fully accounts for earthquakes in general, and for many in particular. What can be imagined greater than a ſhock of the body of the earth? 'Tis greater or leſs in proportion to the ſtate of electrification. And now we can account for ſeveral appearances. In our firſt earthquake, the lord chancellor, maſters in chancery, and ſeveral judges, were ſitting in *Weſtminſter-Hall*, with their back to the wall of the upper end, which is of a vaſt thickneſs. They all relate the ſeverity of the ſhock, from the wall ſeeming to puſh towards them with great violence.

In the earthquake of *September* 30, 1750, Dr. *Stonehouſe*'s dwelling at *Northampton*, the ſtrongeſt in the town, was moſt ſenſibly ſhaken. So it was obſerved likewiſe, that churches were moſt ſubject to its violence. People at divine ſervice felt a great ſhock, which was like ſomewhat, as they imagined, that ruſhed againſt the church wall and roof.

And thus in the earthquake of 1692, *Deal* caſtle, whoſe walls are of immenſe thickneſs and ſtrength, ſhook ſo ſenſibly, that the people living in it, expected it was falling on their heads. And this is the caſe in all earthquakes; the more ſubſtantial the building, the more violent the ſhock is: exactly the mode of electrical vibration.

The city of *Liſbon* is founded on a rock of marble; ſo much the more ſuſceptible of the electric power, which gave it the vibration. Hence the ruins of churches, palaces, houſes, all lie upon their foundations reſpectively; as the houſes of cards made by children, thrown down by a ſlight ſhock of of a table. And so we are to understand of all the rest in other places.

At the same time, that the force of electricity in solids, is as the quantity of matter; we see most evidently by innumerable experiments, that water is equally assistant in strengthening and conveying the force of electricity; and *that* in proportion too to its quantity. And hence is to be deduced the reason of my observation; that the most frequent and dreadful earthquakes have fallen upon maritime places.

In the dreadful catastrophe at *Port-Royal*, 'tis notorious that its violence was chiefly near the sea. So *Lima* could not suffer without its port of *Callao*. *Lisbon*, and the whole *Atlantic* coast is yet a more tremendous and recent example.

That maritime places are most subject, is a strong argument in favour of electricity; when both the solid earth, and the quantity of water concur, to make the shock; exactly as in electrical experiments, when the bottle of water is held in the hand.

Thus, when our mind is discharged of the prejudices of former notions, we discern that every appearance favours the principles we go upon; That subterraneous explosions, could they pervade, and traverse the earth at pleasure, must at last burst and disperse every thing in their way. Yet 'tis not possible for us to imagine, such a kind of vibrations should follow, either by sea or land, as that we are treating of. But electricity compleatly answers it. This accounts for that superficial movement of the earth, that universal instan-

taneous

taneous ſhock, which made every houſe in *London* to tremble, none to fall; that quivering, tremulous, horizontal vibration, highly different from any motion we muſt conceive to be produced from ſubterraneous evaporations. Hence authors tell us, *December* 30, 1739, deſcribing an earthquake in the weſt-riding of *Yorkſhire*, it ſeemed as if the earth moved backward and forward horizontally; and quivering, with reciprocal vibrations.

From electric vibration only can we account for our tenth poſition, of ſprings and fountains being no ways damaged by earthquakes: the motion goes no deeper into the earth, than the force and quantity of the ſhock reaches; which generally is not far; yet it proceeds lower down when the ready paſſage of a well offers, and *there* affects the water contained in it; puts it into an inteſtine vibration, ſo far as to foul it, and raiſe mud from the bottom.

It may ſeem difficult to conceive, how a large portion of the earth's ſurface ſhould be thus capable of electrification. This difficulty is leſſened by reflecting on the nature of electricity, and of the electrical, ethereal fluid, pervading all things: how it is excited by the little motion of a ſmall revolving glaſs globe. By this we electrify the moſt ſolid bodies, to the greateſt diſtance, and with a velocity equal to that of lightning.

We muſt conceive, that when the electric ſhock is communicated to one part of the earth, it extends itſelf proportionably to the force of the ſhock, and to the quantity of electrified ſurface;

and

and to the quality of the matter more or less susceptible of it, more or less apt to propagate it.

Set 1000 men in a row; let every one communicate with those next him, by an iron wire held in their hands: On an electrical shock they all feel alike, at the same instant; and this gives a very good idea of the earthquake.

When the earth is broken up in any large degree 'tis by the sea side, where sometimes on a bold shore, whole streets tumble into the sea or into the gaping earth, now falling into the sea, as the key and custom-house at *Lisbon*: sometimes on a flat and sandy shore, whole streets are rolled along the level into the sea.

I am not sensible of any real objection against our hypothesis. As to the eleventh of my positions or circumstances; it seems true that earthquakes are more frequent in *Italy*, near *Vesuvius*, and by *Ætna* in *Sicily*: And the cause seems apparently owing to those volcanoes, but not so from true reason. This has given the great prejudice to the judgment of the curious, even at this day: But consider the matter impartially, and it will appear so far from being a strong argument in favour of subterraneous eruptions, that it ought to be esteemed a convincing proof of the contrary, and most cogent in favour of my principle. We have but these two or three volcanoes on one quarter of the globe, and two of them toward the warmer climate of it; whereas earthquakes are innumerable, especially in warmer climes. That there are no volcanoes, no discharges of fire and smoke for a continuance, and abundance, after earthquakes; no suspicions of it either from sight or smell, as we know by innumerable examples, as well as in our own country, and experience; is demonstration, that this is not the cause. If the volcanoes were the real cause of earthquakes, we ought assuredly to expect, that in the countries thereabouts, the earthquakes ought to be far more extensive than those in other countries, where are no volcanoes; but this is altogether contrary to experience. For, as the celebrated naturalist *Buffon* observes, such are not extensive, as are near *Ætna* and *Vesuvius*. He further adds, speaking, among many others, of a volcano in the island of *Ternate* [e], "that this burning gulph is less agitated "when the air is calm, and the season mild, than "in storms and hurricanes: (adding) this con-"firms what I have said in my foregoing dis-"course; and it seems evidently to prove, that "the fire which makes volcanoes, comes not "from the bottoms of mountains, but from the "tops, or at least from a very little depth; and "that the hearth (or floor) of the fire is not far "from the summit of the volcanoes; for if this "was not the case, great winds could not contri-"bute to their conflagration." And this in general is a corroborative proof of my whole hypothesis. For there can be no great fire in the earth, where there is no great conveyance of air.

We have one volcano in the cold region of *Iceland*; there is sometimes an earthquake there: but in the countries of that northern latitude, and those of lesser, 'tis obvious in all history, that

[e] Histoire Naturelle. tom. 1. p. 508.

earth-

earthquakes are less frequent than in the more southern. Therefore 'tis easy, and very natural to conclude, from all considerations weighed together, that these volcanoes help to put the earth about them, into that vibratory state and condition of electricity, which is the requisite in my hypothesis; and by that means only, promote a frequency of earthquakes there.

In so surprizing an effect as an earthquake, and so unaccountable a cause as electricity, a cause but recently considered, or known, is it to be wondered at, if some difficulties occur? can we yet pretend to unriddle all the secrets of electricity, though we know some; and in my apprehension are sufficiently clear as to the efficient before us?

Some objections there are, not insuperable. For instance; in electrical experiments the shock is single, and momentary; but earthquakes are felt for some few minutes.

To answer which, we need not urge how fear and frights multiply and magnify objects and appearances: but suppose the vibrations last two minutes, there can be no comparison between our little apparatus in experiments, and the stroke upon seven hills, whereon *Lisbon* was built. The vibrations of musical strings are in proportion to their length, and thickness: the same of bells, and the like. There is no comparison between the snap in our little experiments, and a shock upon the globe of the earth; whence the horrible noise rolling from one eminence to another: as in the air the thunder is re-echoed from one cloud to another.

Again,

Again, ſome find difficulties from apertures in the earth, and ſinkings into the ſea, as is the caſe of the key of *Liſbon*, and the like: So as to mountains opening, and rivers of water guſhing out. I profeſs theſe inſtances move me not in the leaſt, to derive them from the bowels of the earth. The electrical ſtroke from the atmoſphere muſt divide a key, and puſh it into the ſea, or a ſtreet that ſtands on a cliff; as it divides rocks, cliffs, mountains; and tumbles them down, as in the caſe of *Whiteſtone* cliff *Yorkſhire*, in 1755.

Some may object, that if the earth was electrified on an earthquake, every perſon ought to feel it; as when touched in the electrical experiments. But we know, the perſons in a room where ſuch experiments are tried, are not neceſſarily electrified. Yet we find in earthquakes, in fact, many affected, as if electrified, with ſickneſs: And all kind of animals are fully ſenſible of it, and extremely diſturbed.

Sometimes the caſe of *Herculaneum*, and ſuch fancied accidents are quoted, as places ſunk by force of an earthquake. But this is an erroneous poſition. The city remains entire. It was not ſhaken in its laſt cataſtrophe, but buried in *lava* poured upon it from mount *Veſuvius*. Theſe, and ſuch like, are little objections, which it is not worth while to be elaborate in anſwering; as having no foundation on principles of philoſophy.

May 1, 1753, I received a letter from *Peterborough*, by order of the literary ſociety there, with an account of a woman at *Sutton* by *Wansford*, who had been quite deaf for two years laſt paſt, but was perfectly reſtored to her hearing on *Sunday September* 30 preceding, being the day of the earthquake there. She found herſelf reſtored half an hour before the ſhock.

April 1751, we had an account from *Edinburgh* of a perſon reſtored to the uſe of ſpeech, from a reſolution of the nerves, by electrifying. His name *Robert Mowbray*. Theſe and many like caſes confirm our reaſoning.

Though the power that produces theſe amazing ſtruggles in the elements, be manifeſtly one, and the ſame; yet it admits of ſome difference in its action; not only as it may be more or leſs forcible, of more or leſs extent, or as to the different object of its action, but likewiſe in its manner. And this points out ſome names of diſtinction, which are at leaſt uſeful, in all parts of learning and philoſophy.

1. We may therefore denominate one of theſe appearances, the *air-quake*. This ſhews itſelf only in the air, in a moſt horrible rumbling noiſe, like many cannon let off, echoing from one hill to another. It may be called terreſtrial thunder The earth feels not its force, or but ſlightly, or partially, here and there; muſt not be in a proper electric ſtate, and therefore not fully ſuſceptible of the ſhock. This is owing to a preceding rainy, moiſt ſeaſon: which is always adverſe to electricity.

A loud clap of thunder in the atmoſphere, may be heard over a circle of 500 miles diameter. The

The ſame clap diſcharged at the ſurface of the earth the 1ſt of *Auguſt* laſt (1755) was heard all over the counties of *Lincoln* and *Rutland*, and part of all the circumjacent counties. It aroſe to an earthquake, wherever the ground was in a proper diſpoſition for it more or leſs. They that can ſuppoſe this phenomenon to ariſe from under-ground, are not to be argued with.

2. A ſecond kind we may call a *water-quake*; which exerts itſelf in the air and water, as this on *November* 1, with us; cauſing a moſt vehement agitation of that element, lifting it up, and throwing it down by pulſes, toſſing it over the banks of canals, whirling about ſhips and boats, ſhaking, and daſhing them one againſt another, ſtirring up the water from the very bottom, raiſing it from the bottom of wells, and the like.

This appearance occurs in the middle of the ocean: on the land here and there, accompanied with real ſhocks of an earthquake, wherever the earth is in an electric ſtate. This phenomenon muſt acknowledge the ſame aerial origin.

The water is a ready object of its force, both from its mobility, and from its ſolidity. It choſes to run up rivers, to any length; to run along the ſhores, as ready conductors, according to our laſt poſition. They that can fancy ſubterraneous boilings, like a culinary fire, under all the canals, lakes, rivers, ſea-ſhores, and the ocean, affected at the ſame time, over a quarter of the globe, eſpecially in the vessels of water prepared for brewing, are not to be argued with.

We are to conceive, that the electric power falls furiously upon water, by reason of the extreme solidity of the component particles of that most wonderful fluid element: whose sole property it is, of all matter, to be absolutely incompressible. Hence it more readily attracts, and assists the vehemence of the elemental, electric fire. Hence it so readily falls on rocks, mountains, steel, solid buildings, metals, the bones, and joints of animals, and whatever is of the most specific gravity.

This therefore causes a thump at the bottom of a ship at sea, as if striking on the ground; this shakes, and quivers the masts, like an aspen leaf.

3. The third diversity we call properly the *earthquakes*: a tremor of the surface of the earth, accompanied with the two preceding, especially the first, the rumbling noise. These undulations are boundless, as to space, time, or violence, as far as the earth is prepared to receive them. For if a musical string be not rightly stretched, it has no tone. So a wire, in electrical experiments, never so far extended, receives the touch, through its whole length.

It cannot be hard to observe, that all considerations shew the impossibility of a fire underground, perpetrating these dire calamities of earthquakes. The like as to the agitation in the waters, which was perceived even in great vessels of water for brewing: and more, even in lead when in fuſion, at that ſame inſtant of time, as I was credibly informed.

We muſt likewiſe affirm, that the fire and ſmoak of volcano's, is the effect of the electric ſtroke, not the cauſe. The great noiſe is preluſive of the fire, that kindles their component ſulphurs, at the very ſummit; like a match of brimſtone ſtruck by a flint and ſteel. Nor can there be any fire, low in the earth, where there is no conveyance of air, no more than in an exhauſted receiver. And though fires are found in the bottom of coal-mines, and the like, where the air can deſcend; yet we never hear of earthquakes cauſed by them.

4. A fourth kind, I hold to be what we vulgarly call a *water-ſpout*, which is ſeen both on the water, and on the land. 'Tis a partial exerciſe of the aereal power, that lifts up the water in the ocean, rivers, wells, canals. A ſingle vortex or column, ſometimes viſible, of a great height.

In the accounts from *Cadiz* and other places, the water is ſeen coming from the great ocean, like a mountain, and when at the ſhore, covering the land: and many of theſe like columns or ridges 50 or 60 foot high, more or leſs, ſucceeding one another. The like appearance, *cæteris paribus*, in lakes, canals. All theſe are owing to the ſame aereal power that makes the water-ſpouts.

These four kinds proceed all from the same cause, under some different circumstances, single or complex, greater or lesser. The *rationale* of them we leave to further disquisition, content to point out some of them, and enumerate their species.

We have seen universally that earthquakes and agitations happen in a serene sky. We have asserted their cause to be electrical strokes from the atmosphere, the same as thunder and lightning. Now that thunder and lightning which produces earthquakes, is found in a clear sky, free from clouds, was known to that great genius *Horace*, as appears very fairly from Ode XXXIV. of his first book; but not commonly understood, from want of a true pointing. Thus,

——Namque Diespiter
Igni corusco nubila dividens
Plerumque; per purum, tonantes
Egit equos, volerremque currum.
Quo bruta tellus, et vaga flumina,
Quo Styx et invisi horrida Tænari
Sedes, Atlanteusque finis
Concutitur.

A comma is usually put after the word *dividens*, but erroneously. Mr *Baxter* discerned it ought to be after the word *plerumque*, otherwise 'tis not agreeable to that good sense we ought to find in our poet: and it now shews that he was a philosopher too.

It

It may be thus tranſlated, and accommodated to the preſent times.

——For hitherto great *Jove*,
Who o'er the clouds his thundring chariot drove:
Of late his fierceſt lightning has been ſeen
To dart impetuous thro' the ſky ſerene.
The ſolid earth an awful tremor feels,
The rivers dance before his chariot wheels:
To *Afric*'s ſhores the rapid ſhock extends,
E'en to the dreadful *Stygian* cave deſcends;
The yawning realm of *Tænarus* appears,
Awakens conſcience with unuſual fears.

PHAEOMENA

OF THE

Great EARTHQUAKE

Of *November* 1, 1755, in various parts of the Globe.

EUROPE.

In GREAT-BRITAIN and IRELAND.

BARDFIELD, *Essex*. The waters in ponds greatly agitated between 11 and 12 in the morning.

BARLBOROUGH, *Derbyshire*. Between 11 and 12, in a boat house on the west side of a large body of water, called *Pibley Dam*, suppos'd to cover at least thirty acres of land, was heard a surprizing and terrible noise, and a large swell of water came in a current from the south, and rose two feet on the sloped dam head at the north end of the water. It then subsided, but returned again immediately, though with less violence. The wa-

ter

ter continued thus agitated for three quarters of an hour, but the current grew weaker and weaker, till at laſt it entirely ceaſed. During this diſturbance, not a breeze of wind was heard, nor a wave ſeen upon the ſurface. A hardy young fellow was ſent to the boat-houſe to ſee if any beaſt was there plunged in the water, but was ſo ſhock'd with the noiſe, and by the boats tumbling about and beating againſt the ſides of the houſe, that when he returned he was not able, at firſt, to give a rational anſwer to any queſtion that was aſked him. When all was ſtill and quiet, it appeared by a ſtake which had been drove down in the pond when the boat-houſe was built, that the water on that ſpot had riſen about eight inches.

BOCKING, *Eſſex.* The ſame as at *Bardfield.*

BUSBRIDGE, *Surrey.* (near *Godalmin*) At half an hour after ten in the morning, the weather being remarkably ſtill, without the leaſt wind, in a canal near 700 feet long, and 58 feet broad, with a ſmall ſpring conſtantly running through it, a very unuſual noiſe was heard at the eaſt end, and the water there was obſerved to be in great agitation; it raiſing it ſelf in a heap or ridge in the middle, which extended length-wiſe about 30 yards, and between two and three feet above the uſual level: After which the ridge heeled or vibrated towards the north ſide of the canal, with great force, and flowed above eight feet over the graſs walk on that ſide. On its return back into the canal, it again ridged in the middle, and then heeled, with yet greater force, to the ſouth ſide, and flowed over its graſs walk; during which

latter

latter motion, the bottom on the north ſide was left bare of water for ſeveral feet wide. The water being returned a ſecond time into the canal, the heelings grew leſs and leſs, yet ſo ſtrong as to make it flow ſeveral times more over the ſouth bank, which is ſomething lower than the other. In about a quarter of an hour after the firſt appearance, the water became quiet and ſmooth as before. During the whole time there was a great perturbation of the ſand from the bottom, with a noiſe like to that of water turning a mill. The higheſt part of the walk, over which the water flowed, was about 20 inches above the water level. No motion was taken notice of in the water at the weſt end of the canal.

CAVERSHAM, *Oxfordſhire*. (near *Reading*) People were alarmed with a very great noiſe about 11 in the morning, as if part of their houſe had been falling down: Upon examination however it did not appear, that the houſe was at all damaged; but a vine which grew againſt it was broken off, and two dwarf trees were ſplit.

COBHAM, *Surrey*. Between 10 and 11, a perſon was watering a horſe in hand, at a pond fed by ſprings, which had no current. Whilſt the horſe was drinking, the water ran away from him, and moved towards the ſouth with ſwiftneſs, and in ſuch a quantity, as left the bottom of the pond bare; then returned with that impetuoſity, which made the man leap backwards, to ſecure himſelf from its ſudden approach. It went back again to the ſouth, with a great ſwell, and returned again. Its riſe was above a foot. The ducks were

were alarmed at the firſt agitation, and flew all inſtantly out of the pond. There was a particular calm all this time.

Conistone-water, *Cumberland,* (a lake about five miles in length) A ferry man ſtanding at their landing place, as he gueſſes about 10 in the morning, was ſurprized to ſee the water flow above a yard upon the bank when there was not the leaſt wind, and the water quite calm; and continued its motion backwards and forwards about five minutes. The perpendicular riſe might be about a foot.

Cranbrook, *Kent.* The people were very much alarmed and fancied they felt an earthquake. The waters of ſeveral ponds, in this and the adjacent pariſhes, were in ſuch motion, that they overflowed their banks, and then returned back, and overflowed the other ſide.

Cork, *Ireland.* At 36 minutes after 9 two ſhocks of an earthquake were felt at about half a minute's interval: The limits of the places affected were, ſouthward, *Watergate-lane*, *Chriſt-church-lane*, and *Playhouſe-ſtreet*; northward, *Broad-lane*, *Coal-Quay*, and *Draw-bridge.*

Creston-ferry, *Devon.* (a mile ſouth-eaſt of *Plymouth*) About 4 in the afternoon, almoſt immediately after high-water, the tide made a very extraordinary out, or receſs, and left two laden paſſage-boats, at once, quite dry in the mud, though they were, a minute or two before, in four or five feet water. In leſs than eight minutes the tide returned with the utmoſt rapidity, and floated both the boats again, ſo that they had

ſix

ſix feet water. The ſea ſunk and ſwelled, tho' in a much leſs degree, for near half an hour longer, and at the next morning's tide there, ſeveral very large ſurges, which drove ſhips from their moorings, broke ſome of the hawſers, and twirled veſſels about in a very odd manner.

Crunill-passage, *Devon.* Over an arm of the ſea, about two miles weſt of *Plymouth*, the ſame phænomena were obſerved, as at *Creſton-Ferry.*

Dunstall, *Suffolk.* (near *Bury*) The water of a pond roſe gradually, for ſeveral minutes, in the form of a pyramid, and fell down like a water-ſpout; whereas other ponds thereabouts had a ſmooth flux and reflux, from one end to the other.

Durham *city.* (near it) About half an hour after 10, a gardener was alarmed by a ſudden ruſhing noiſe from a pond; as if the head of the pond had broken down: when caſting his eye on the water, he ſaw it gradually riſe up, without any fluctuating motion, till it reached a grate, which ſtood ſome inches higher than the common water-level, thro' which it diſcharged itſelf for a few ſeconds. Then it ſubſided as much below the mark it roſe from, as it was above it in the greateſt elevation, and continued thus riſing and falling about ſix or ſeven minutes, making four or five returns in about one. The water ſtill continued to have ſome commotion, but it was nothing conſiderable. The ebb and flow were each about half a foot in the perpendicular. The pond is about 40 yards long, and 10 broad.

Early-court, *Berks.* (near *Reading*) About 11 o'clock, a gardener ſtanding by a fiſh-pond, felt

felt *a moſt violent trembling of the earth*, which laſted upwards of fifty ſeconds: Immediately after which, he obſerved a motion of the water, from the ſouth end of the pond to the north end, leaving the bottom on the ſouth end altogether without water, for the ſpace of ſix feet. It then returned, and flowed at the ſouth end, ſo as to riſe three feet up to the ſlope bank, and immediately went back again to the north, where it likewiſe flowed three feet up the bank: And in the time between the flux and reflux, the water ſwelled up in the middle of the pond, collected in a ridge, about 20 inches higher than the level on each ſide, and boiled like a pot. This agitation from ſouth to north, and from north to ſouth alternately, laſted about four minutes; and there ſeemed to be little or no motion in the direction of eaſt and weſt, the weather being perfectly calm during the whole time.

EASTHWAITE-WATER, *Cumberland.* (a lake about a mile and a half in length, near *Hawkeſhead)* A like agitation, though in a leſs degree, and ſhorter continuance, as at *Coniſtone-water*, and at the ſame time.

EATON-BRIDGE, *Kent.* In a pond, about an acre in ſize, a dead calm, and no wind ſtirring, ſome perſons heard a noiſe, and imagining ſomething had tumbled in, ran to ſee what was the matter; when, to their ſurprize, they ſaw the water, open in the middle, ſo that they could ſee a poſt a good way down, almoſt to the bottom, and the water daſhing up over a bank about two

foot

foot high, and perpendicular to the pond. This it did ſeveral times, making a great noiſe.

ENFIELD, *Middleſex.* Agitations on the water.

EYAM-EDGE, *Derbyſhire.* (in the Peak) The overſeer of the lead mines, ſitting in his writing room, felt, about 11 o'clock, one ſhock, which very ſenſibly raiſed him up in his chair, and cauſed ſeveral pieces of lime or plaiſter to drop from the ſides of the room. The roof of it was ſo violently ſhook, that he imagined nothing leſs than the engine ſhaft was run in; whereupon he immediately went to ſee, and found the ſhaft open, and all things about the ſpot in their proper order. In the morning, coming through a field about 300 yards from the mines, there was nothing uncommon to be ſeen, but in his return at evening he obſerved a cleft about one foot deep and ſix inches over; its continuation from one end to the other, was near 150 yards, being parallel to the range of the vein on the north ſide. Theſe were the moſt remarkable circumſtances which happened on the ſurface of the earth.

Two miners at the aforeſaid time were employed carting, or drawing along the drifts, the ore and other minerals to be raiſed up the ſhafts. The drift, in which they were working, is about 120 yards deep, and the ſpace from one end to the other 50 yards, or upwards. He at the end of the drift had juſt loaded his cart, and was drawing it along, but was ſuddenly ſurprized by a ſhock, which ſo terrified him, that he immediately quitted his employment, and ran to the weſt end of the drift to his partner, who was not

leſs

leſs terrified than himſelf. They durſt not attempt to climb the ſhaft, leſt that ſhould be running in upon them, but conſulted what means to take for their ſafety. Mean while they were alarmed by a ſhock much more violent than the former; which put them in ſuch a conſternation, that they both ran precipitately to the other end of the drift. There was another miner working at the eaſt end of the vein, about 12 yards below their level, who called out to them, imagining they were in danger of being killed by the ſhafts running in upon them, which he ſuppoſed was the caſe; and told them, if by any means they could get down the ſhaft to him, they would be more ſecure, becauſe the cavity, where he was working, was encompaſſed with ſolid rock. They went down to him, where after obſerving they had neither of them received any misfortune, he told them that the violence of the ſecond ſhock had been ſo great, that it cauſed the rocks to grind one upon another. His account was interrupted by a third ſhock, which, after an interval of four or five minutes, was ſucceeded by a fourth; and about the ſame ſpace of time after, by a fifth; none of which were ſo violent as the ſecond. They heard after every ſkock a loud rumbling in the bowels of the earth, which continued about half a minute, gradually decreaſing, or ſeeming at a greater diſtance. They imagined, that the whole ſpace of time, from the firſt ſhock to the laſt, might be about 20 minutes; and they tarried about ten minutes in the mine, after the laſt ſhock. As they went along the drifts, they obſerved, that ſeveral pieces

pieces of minerals were dropped from the ſides and roof, but all the ſhafts remained entire, without the leaſt diſcompoſure: The ſpace of ground at the aforeſaid mines, wherein it was felt, was 960 yards, which was all that was at that time in workmanſhip.

FINCHINGFIELD, *Eſſex*. Between 11 and 12 the water of a pond, which has no communication with any river, ran up hill into a ditch. Juſt before the agitation of the water, the geeſe in the pond ſcreamed vehemently.

FRAMLINGHAM, *Suffolk*. (near *Ipſwich*) A large pond was greatly agitated.

GAINSBOROUGH, *Lincolnſhire*. The water in this port roſe five or ſix feet, and fell again in a minute or two.

GUAVA'S-LAKE, *Cornwal*. A ketch of war veered round upon her anchors, keeping her head by turns to the flux and reflux; and in the decline of the commotion they hove the log to eſtimate the velocity of the water, and found it to run at the rate of ſeven miles in an hour.

GUILFORD, *Surry*. (near it) In a mill-pond, a great ſwell and agitation of the water was obſerved by a perſon who ſtood over it all the while, on a bridge; and in a back ſtream it was very conſiderable, and came with violence againſt the bank, but no ſenſible reflux was obſerved.

HEYLE, *Cornwal*. (a little harbour about four miles north of the *Mount*, on the *Severn* ſea) The agitation did not make its appearance, till an hour or little more after the ebb began, or about 4 in the afternoon, which is eaſily accounted for (ſays

(ſays the obſerver) by the circuit of land at the extremity of the county, which the ſwell muſt have made before it could reach into the north chanel to St. *Ives* and *Heyle*. In this inland half-tide harbour it continued viſible but an hour and half; the greateſt flux was about the middle of that time, the ſurge being then ſeven feet high; but in general it roſe and fell but two feet only; owing, as he ſuppoſes, to the force and quantity of water being broke in its advances into ſo retired a creek.

HORSMANDEN, *Kent*. The ſame phænomena as at *Cranbrook*.

HULL, *Yorkſhire*. The ſame as at *Gainſborough*.

HUNSTON, *Norfolk*. Two gentlemen went out a ſhooting on the ſea-ſhore, and were in great danger of being drowned by the ſea's ſudden flowing before its uſual time.

St. IVES, *Cornwal*. (at the peer) The water roſe between eight and nine feet, and floated two veſſels, before quite dry, but all ſmooth; no ſea broke.

KINSALE, *Ireland*. Between the hours of two and three in the afternoon, the weather being very calm, and the tide near full, a large body of water ſuddenly poured into the harbour, with ſuch rapidity, that it broke the cables of two ſloops, each moor'd with two anchors, and of ſeveral boats lying between *Scilly* and the town; which were carried up, then down the harbour, with a velocity far exceeding that of a ſhip or boat, though favoured with all the advantageous circumſtances of tide and wind, in any degree of violence: But juſt at the time that univerſal miſchief was thought unavoidable by all the veſſels running fowl of one another, an eddy whirled them round ſeveral times, and hurried them back again with the ſame rapidity. This was ſeveral times repeated; and while the current ruſhed up at one ſide of the harbour, it poured down with equal violence at the other. A veſſel that lay all this time in the pool, did not ſeem to be any ways affected; nor was the violence of the currents much perceived in the deeper parts of the harbour, but raged with moſt violence on the flats. The bottom of the harbour, which is all a ſlab, was much altered, the mud being waſhed from ſome places, and depoſited in others. The perpendicular riſe of the water at one quay was meaſured five feet and a half, and is ſaid to have been much higher at the market quay, which it overflowed and powered into the market-place with ſuch rapidity, that ſome who were on the quay, immediately ran off, on the firſt riſe of the water, but could not do it with expedition enough to prevent their being overtaken, and up to the knees. The agitations of the water were communicated ſome miles up the river, but, as in the harbour, were moſt perceivable in the ſhalloweſt places. The ſucceſſive riſings and fallings of the water ſeemed to continue about ten minutes, and then the tide returned to its natural courſe. Between 6 and 7 in the evening the water roſe again, though not with ſo great violence as before; and it continued alternate ebbs and flows till 3 in the morning. The

waters

waters did not riſe gradually at firſt, but with a hollow and horrid noiſe ruſhed in like a deluge, roſe ſix or ſeven feet in a minute, and as ſuddenly ſubſided: It was as thick as puddle, very black, and ſtank intolerably. By different accounts the water was affected in the ſame manner, all along the coaſt, to the weſtward of this harbour.

LANDS-END, *Cornwal.* The commotion of the waters was perceived there.

LEE, *Surrey.* (in *Whitley* pariſh) A canal or pond was ſo violently agitated, that the gardener, on the firſt appearance, ran for help, thinking a number of otters were under the water deſtroying the fiſh.

LOCH-KETERIN, *Scotland.* Agitated at the ſame time as *Loch-Lommond*, which ſee.

LOCH-LOMMOND, *Scotland.* At half an hour paſt nine in the morning, all of a ſudden, and without the leaſt guſt of wind, the water roſe againſt its banks with great rapidity, but immediately retired, and in five minutes time ſubſided, till it was as low in appearance, as any body then preſent had ever ſeen it in the greateſt ſummer drought; and then it inſtantly returned towards the ſhore, and in five minutes time roſe again as high as it was before. The agitation continued at this rate till fifteen minutes after ten the ſame morning, taking five minutes to riſe, and as many to ſubſide; and from fifteen minutes after ten, till eleven, every riſe came ſomewhat ſhort in height of that immediately preceding, taking five minutes to flow, and five to ebb, untill the water was ſettled, as it was before the agitation. The height to which the loch rofe perpendicular, was meafured and found to be two feet, four inches.

LOCH-LONG, *Scotland.* Agitated at the fame time as *Loch-Lommond*, which fee.

LOCH NESS, *Scotland.* At half an hour after nine, a very great agitation of the waters was feen by divers perfons; and about ten the river *Oich*, which lies on the north fide of fort *Auguftus*, and runs from weft, into the head of the loch, was obferved to fwell very much, and run upwards, from the loch, with a pretty high wave, about two or three feet higher than the ordinary furface, with a pretty quick motion againft the wind, and a rapid ftream, about two hundred yards up the river; then broke on a fhallow, and flowed about three or four feet on the banks, on the north fide of the river, and returned again gently to the loch. It continued ebbing and flowing in that manner for about an hour, without any waves fo remarkable as the firft, till about eleven o'clock, when a wave higher than any of the reft came up the river, and, to the great furprize of all the fpectators, broke with fo much force on the low ground, on the north fide of the river, as to run upon the grafs upwards of thirty feet from the river's bank. *Loch-nefs* is about twenty miles in length, and from one to one and a half mile broad; bears from fouth-weft to north-eaft. It is vaftly deep, its foundings in many places being from one hundred, to an hundred and thirty-five fathom, which is greatly below the level of the fea at *Invernefs.* Its fides are moft part rocky, and it deepens immediately from them. About three mufketmuſket-ſhot from the river *Oich* it meaſures about one hundred and twenty fathom in depth. There was no extraordinary muddineſs obſerved in the water, upon this occaſion, though it did not appear quite ſo clear as uſual. The morning was cold and gloomy, and a pretty briſk gale of wind blowed from weſt-ſouth-weſt.

LUTON, *Bedfordſhire.* The water of a pond was ſtrongly agitated, and ſeveral times overflowed its banks on one ſide, and ebbed ſix feet on the other; this was between ten and eleven in the morning.

MEDHURST, *Suſſex.* In a mill-pond, the ſwell of the water, rolling towards the mill, was ſo remarkable, that the miller imagined a ſluice had been opened at the upper end of the pond, and had let a back water into it; but upon ſearch it was found ſhut as uſual: Upon its retreat, it left ſome fiſhes upon dry land. Below the mill the ſwell of the water was ſo great, as to drive the ſtream upwards, back into the conduit of the mill. The pond in lord *Montacute*'s park, in the neighbourhood, was likewiſe greatly agitated at the ſame time.

MOUNTS-BAY, *Cornwal.* A little after two in the afternoon, the weather fair and calm, as it had been for ſix days before, the barometer unuſually high, the thermometer about temperate, and the little wind there was, at north-eaſt, there happened here, and the parts adjacent, the moſt uncommon and violent agitation of the ſea ever remembered. About half an hour after ebb, the ſea was obſerved at the *Mount-pier* to advance ſuddenly from the eaſtward. It continued to ſwell and riſe for the ſpace of ten minutes; when it began to retire, running to the weſt, and ſouth-weſt, with a rapidity equal to that of a mill-ſtream deſcending to an underſhot wheel: It ran ſo for ten minutes, till the water was ſix feet lower, than when it began to retire. The ſea then began to return, and in ten minutes it was at the before-mentioned extraordinary height: In ten minutes more it was ſunk as before, and ſo it continued alternately to riſe and fall between five and ſix feet, in the ſame ſpace of time. The firſt and ſecond fluxes and refluxes were not ſo violent at the *Mount-pier* as the third and fourth, when the ſea was rapid beyond expreſſion, and the alterations continued in their full fury for two hours: They then grew fainter gradually, and the whole commotion ceaſed about low water, five hours and a half after it began. At the mount the fiſhermen got to their boats, then riding off the pier, as ſoon as the commotion was obſerved, concluding that a violent ſtorm was at hand: They were no ſooner on board, than their boats were heaved in with the ſurf; but they were no ſooner in the pier, and ſtruggling to ſecure themſelves and boats, as much as their aſtoniſhment would permit, than their boats were hurried back again, through the gap or mouth of the pier, with incredible velocity: When they had gone off as far as the reflux determined, they were carried in, and out again, with an impetuoſity, which no ropes could withſtand, and which would have deſtroyed both men and boats immediately, if in

their

their paſſage, they had touched the leaſt ſtone of the pier. What preſerved them, was not the rudder, or the oar, but the ſame ſtream and current, which put them in danger; for it had neither in or out-let but through that narrow gap, and therefore ſet in directly, and out.

MOUSHOLE-PIER, *Cornwal.* (in *Mounts-bay*) The agitations of the ſea did not materially differ from thoſe at *Newlin-pier*.

NETTLEBED, *Oxfordſhire*. A reſervoir there called *Wombone-pond*, was found quite empty of water, the bottom having ſunk within the earth and left an unfathomable cavity. It had been obſerved to be full at eight o'clock the night before.

NEWLIN-PIER, *Cornwal.* (a mile weſt of *Penzance*) The flux was obſerved to come in from the ſouthward, the eaſtern current (ſays the curious obſerver) being quite ſpent. It was nearly at the ſame time as at *Mounts-bay* and *Penzance*, but in a manner ſomewhat different; it coming on like a ſurge or high creſted wave. The firſt agitations were as violent as any; and after a few advances and retreats, at their greateſt violence, in the ſame ſpace of time as at the *Mount*, the ſea grew gradually quiet, after it had roſe, to the infinite amazement of the ſpectators, ten feet perpendicular at leaſt: This is five feet more than at the *Mount-pier*, and two feet more than at *Penzance*, attributed, by the obſerver, to the angle or creek in which *Newlin* lies; wherein the waters were reſiſted, and accumulated by the ſtraitneſs of the ſhores, and the bent of the weſtern land; whereas at *Penzance* the waters were leſs confined, and confequently could not rife fo high; but at the *Mount* (at that time an ifland) the fea had full room to fpread, and difperfe it felf, and there rofe leaft of all. *See Penzance.*

Oich, river, *Scotland*, greatly agitated. See *Loch-Nefs*.

Peerless-pool, *Middlefex*. (in *Old-ftreet* parifh, near *London*) Between the hours of ten and eleven in the morning, one of the waiters there being engaged with his fellow-fervant, in fome bufinefs, near the wall inclofing the ground which contains the fifh-pond, and accidentally cafting his eye on the water, was furprized to fee it greatly moved, without the leaft apparent caufe, as the air was entirely calm; he called to his companion, who was equally ftruck with the fight of it. Large waves rolled to and from the bank near them, at the eaft end, for fome time, and at laft left the pond dry for feveral feet, and in their reflux overflowed the bank ten or twelve feet, as they did the oppofite one, which was evident from the wetnefs of the ground about it. This motion having continued five or fix minutes, the two waiters ftept to the cold bath near the fifh pond, but no motion was by them obferved in it, nor by a gentleman who had been in it, and was then dreffing, and who went immediately with the waiters to the fifh-pond, and was a third witnefs of the agitation there. When all had ceafed, thefe three went to the pleafure bath, between which and the fifh-pond the cold bath is fituated; they found it then motionlefs, but to have been agitated in the fame manner with the fifh-pond, the water leaving plain marks of its having overflowed the banks, and risen to the bushes on the sides of them.

PENZANCE, *Cornwal.* The pier lies three miles west of the *Mount*, and the reflux was first observed here forty-five minutes after two: The influx came on from the south-east, and south south-east: From whence the observer gathers, that the force, from which the agitation proceeded, lay at south nearly, or south-west of the bay, and the sea reaching first the eastern lands (which project a great deal more than those of the west) was thence reflected, and came upon the *Mount* in an easterly direction: but further on the west this eastern current had lost its strength, and the sea came into *Penzance* from the south-south-east, more directly from the points of its momentum. Here the greatest rise was eight feet, and the greatest violence of the agitation about three o'clock. See *Mounts-bay*.

PLYMOUTH, *Devonshire.* About four in the afternoon, there was an extraordinary *boar*, as the sailors call it. The sea seemed disturbed about twenty minutes before, though there was very little wind that day, or for some days before. The sky seemed that day very cloudy, in the morning very full of little fiery red clouds, in the afternoon very louring, and in many places of a very odd copper-colour; the atmosphere excessively thick and dark, but not a drop of rain fell. The boar drove several ships from their moorings, and broke some of the hawsers.

PONTY-POOL, *Monmouthshire.* (near it) The river *Frood* sunk, by the fall of a rock, into the earth, and is lost; not yet having been discovered to to have broken out any where again, though it may be obſerved to run about ten yards under ground.

Pool, *Dorſetſhire.* Between ten and eleven o'clock in the morning, the ſea at the quay was violently agitated, though calm juſt before: Ships were toſſed and broke from their moorings. Some felt a ſlight earthquake at land.

Portsmouth, *Hants.* About thirty-five minutes after ten in the morning there was obſerved, in the dock-yard, an extraordinary motion of the waters in the north dock, and in the baſon, and at two of the jetty heads. In the north dock whoſe length is two hundred and twenty-nine feet, breadth ſeventy-four feet, and at that time ſeventeen feet and a half depth of water, the *Goſport* man of war of forty guns, was juſt let in to be dock'd, and well ſtayed with guys and hawſers. On a ſudden, the ſhip ran backward near three feet; and, by the libration of the water, the gates alternately opened and ſhut, receding from one another near four inches. In the baſon, whoſe length is about two hundred and forty feet, breadth two hundred and twenty feet, and at that time about ſeventeen feet depth of water, ſhut in by two pair of gates, lay the *Berwick* of ſeventy guns, the *Dover* of forty guns, both in a direction nearly parallel to the *Goſport*, and a merchant ſhip of about ſix hundred tons, unloading of tar, lying in an oblique direction to the others. Theſe ſhips were obſerved to be agitated in like manner with the *Goſport*, and the tar ſhip to roll from ſide to ſide. The ſwell of the waters againſt the ſides of the

the bafon was obferved to be nine inches; one of the work-men meafured it between the librations. The *Naffau*, a feventy gun fhip, lying a long-fide a jetty head, between the north dock and the bafon; alfo the *Duke*, a ninety gun fhip, lying againft the next jetty head, to the fouthward, both in a direction nearly at right angles to the others, were obferved to be rocked in the fame manner, but not quite fo violently. The dock and bafon lie nearly eaft and weft, on the weft fide of the harbour.

ROCHFORD, *Effex*. About ten in the morning, in a pond adjoining to the church-yard, the water was obferved to flow a confiderable way up the mouth of the pond, and then returning, to flow up the oppofite fide, repeating this fort of motion for about three quarters of an hour. At the very time of this fluctuation, two other ponds, which are but a fmall diftance from the former, were remarked to be ftill and quiet. The motion of the water in the firft pond was only from eaft to weft, and from weft to eaft, alternately. This pond is very large, and almoft round: Its mouth is on the eaft fide. The two neighbouring ponds lie in length from north to fouth, and are comparatively very narrow in their breadth from eaft to weft.

SHIRBURN-CASTLE, *Oxfordfhire*. At a little after ten in the morning, a very ftrange motion was perceived in the water of a moat which incompaffes the houfe. There was a pretty thick fog, not a breath of air, and the furface of the water all over the moat as fmooth as a looking-glafs, glaſs, except at one corner, where it flowed into the ſhore, and retired again ſucceſſively, in a ſurprizing manner. How long it had done ſo before, or in what manner it began to move, is uncertain; the flux and reflux, when ſeen, were quite regular. Every flood began gently; its velocity increaſed by degrees, till at laſt, with great impetuoſity, it ruſhed in till it had reached its full height, at which it remained for a little while, and then again retired, at firſt gently ebbing, at laſt ſinking away with ſuch quickneſs, that it left a conſiderable quantity of water entangled amongſt the pebbles, laid to defend the bank, which ran thence in little ſtreams over the ſhore, now deſerted by the water, which at other times always covers it. As the ſlope of the ſides of the moat is very gentle, the ſpace left by the water at its reflux was conſiderable, though the difference between the higheſt flood and loweſt ebb of theſe little tides, if the expreſſion may be allowed, was but about four inches and an half perpendicular height; the whole body of water ſeeming to be violently thrown againſt the bank, and then retiring again, while the ſurface of the whole moat, all the time, continued quite ſmooth, without even the leaſt wrinkle of a wave. The time it took up in one flux and reflux, as it was not then obſerved, cannot be gueſſed at. Several pieces of white paper lay at the bottom of the water, about four foot deep: Theſe could be perceived to move backward and forwards, keeping pace with ſome weeds, and other things, which floated on the top of the water, as it ebb'd and

and flow'd. Lord viscount *Parker*, who had observed these reciprocations, being desirous to know, whether the motion was universal over the moat, sent a person to the other corner of it, at the same end that himself stood, and about twenty-five yards from him, to examine whether the water moved there, or not. He could perceive no motion there, or hardly any: But another, who went to the north-east corner of the moat, diagonally opposite to his lordship, found it as considerable as where he was. His lordship imagining, that in all probability the water at the corner diagonally opposite to where he was, would sink, as that by him rose, he ordered the person to signify, by calling out, when the water by him began to sink, and when to rise. This he did, but to his lordship's great surprize, he found, that, immediately after the water began to rise at his own end, he heard his voice calling that it began to rise with him; and in the same manner heard that it was sinking at his end, soon after he perceived it to sink by himself. They might be about ninety or an hundred yards asunder. His lordship sent a person to a pond just below where himself stood, who called to him in the same manner. The water rose and fell in that pond; but though he stood at the south-west corner of that pond, as my lord did at the south-west corner of the moat, it did not rise and fall by him in that pond, at the same time as it rose and fell by his lordship in the moat, but sunk sometimes when the moat rose, and rose when the moat sunk, as it seemed by his calling, the rising and falling seeming to be quicker than in the moat, though but little: He might stand about

about forty yards off. His lordſhip ſent to three other ponds, in all which the agitation was very conſiderable. The ſwells which ſucceeded one another, were not equal, nor did they increaſe or diminiſh gradually; for ſometimes, after a very great ſwell, the next two or three would be ſmall, and then again would come a very large one, followed by one or two more as large, and then leſs again. His lordſhip having ſtood by the moat a good while, went away, and returning again in about half an hour's time, found it perfectly ſtill.

STONEHOUSE LAKE, *Devonſhire.* (communicating with an arm of the ſea) The boar or ſwell came in with ſuch impetuoſity, that it drove every thing before it, tearing up the mud, ſand and banks, and broke a large cable, by which the foot paſſage boat is drawn from ſide to ſide of the lake.

SWANZEY, *Glamorganſhire.* (in *Briſtol* channel) *See White-rock.*

TARFF river, *Scotland.* (ſouth of fort *Auguſtus*) Was agitated at the ſame time and manner as the river *Oich.* See *Lochneſs* and *Oich.*

TENTERDEN, *Kent.* Between 10 and 11 in the morning, ſeveral ponds here and in the neighbourhood were greatly agitated; the water being forced up the banks, with much violence, foaming, fretting, and roaring like the coming in of the tide. Some flowed up three times in this manner, others circled round in eddies, abſorbing leaves, ſticks, &c.

THAMES, river. (at *Rotherithe*) Some perſons being in a barge, unloading timber, between 11 and 12 o'clock, were ſurprized by a ſudden heaving up of the barge, from a ſwell of the water,

not

not unlike what happens when a ſhip is launched from any of the builders yards in the neighbourhood. But the ſtate of the tide did not then ſuit with the launching of ſhips, and they were afterwards certain that no ſhip was launched at that time.

TOPSFIELD, *Eſſex*. The water of a pond roſe very high.

TUNBRIDGE town, *Kent*. The waters agitated.

WHITEHAVEN, *Cumberland*. The waters agitated.

WHITE-ROCK, *Glamorganſhire*. (above a mile above *Swanzey*) About two hours ebb of the tide, and near three quarters after ſix in the evening, a great head of water ruſhed up with a great noiſe, floated two large veſſels, the leaſt of them above two hundred tons, (one whereof was almoſt dry before) broke their moorings, and hove them acroſs the river, and had like to overſet them, by throwing them on the banks. The whole did not laſt ten minutes, the riſe and fall, and what is moſt remarkable, it was not felt in any other part of the river, ſo that it ſhould ſeem to have guſhed out of the earth at that place: For near *Swanzey* town, and mouth of the river, there is a paſſage-boat, that was paſſing at that time, and had been for the whole day, and there nothing was felt of it.

WINDERMERE-WATER, *Cumberland*. (a lake about ten miles long from north to ſouth) About ten o'clock in the forenoon, a fiſhing-boat being drawn aground, one of the men aſhore, and the other ſitting in the boat, the lake quite full, and as ſmooth as glaſs, and not a breath of wind; on a ſudden the water ſwelled, floated the boat, heaved it up about its length farther upon land, and took it back again, in the falling back of the wave.

 This

This flux and reflux continued about eight or ten minutes, gradually decreaſing: Here they heard no remarkable noiſe. Some ferry-men, buſy at the ſame time on ſhore, about the middle part, gave the like account in every particular, only that their boat was moored, and could not be driven on ſhore; the ſwell they judged to be about knee-high above the common ſurface. Some huſbandmen that were at work that forenoon in a field, within-ſight of the lake, about two miles and a half from the foot or ſouth end of it; about ten, heard a noiſe from towards the water, like, as they imagined, the ſound of the ſlate off the whole ſide of any large building, ſliding down the roof at once, and expected it to be ſome ſtrong guſt of wind coming at a diſtance: The water was quite ſtill before and ſmooth, but on that noiſe they obſerved a narrow rippling in the lake, from the point of a rock.

WYMANSDEL-MEARE, *Weſtmoreland.* Was agitated in a very extraordinary manner; for in an inſtant the waters roſe ſeven feet, and again as ſoon ſubſided; ſo that two fiſher-men who were in a boat, near the edge of the lake, found themſelves by one wave carried into it a conſiderable way, and were ſo aſtoniſhed with the ſudden tranſportation, as to declare they imagined that the laſt day was come.

YARMOUTH, *Norfolk.* A little before noon, without any wind ſtirring, the water in the haven was violently agitated, and ſuddenly roſe ſix feet, and the ſhips had an uncommon motion, ſo that the caulkers left off work for ſome time.

In BOHEMIA.

TOPLITZ (a village famous for its medicinal baths, nine *Bohemian* miles north-west of *Prague)* These waters were discovered in the year 762; from which time the principal spring of them had constantly thrown out the hot water in the same quantity, and of the same quality. On *November* 1, 1755, between eleven and twelve in the morning, the chief spring cast forth such a quantity of water, that in the space of half an hour all the baths ran over. About half an hour before this vast increase of the water, the spring grew turbid, and flowed muddy; and having stopped intirely near a minute, broke forth again with prodigious violence, driving before it a considerable quantity of a reddish oker, or *crocus martialis:* After which it became clear, and flowed as pure as before; and continues still to do so; but it supplies more water than usual, and that hotter, and more impregnated with its medicinal quality.

In FRANCE.

ANGOULESME, capital of *Angoumois*; about a league from this city a subterraneous noise, like thunder, was heard, and presently after the earth opened and discharged a torrent of water mixed with red sand. Most of the springs in the neighbourhood sunk, in such a manner, that for ſome time it was thought they were quite dry, and the *Charante* at the ſame time ſunk conſiderably, and then ſwelled up in a ſurprizing manner.

BAYONNE, *Gaſcony*. A pretty ſmart ſhock was felt.

BLEVILLE, *Normandy*. (a league from *Havre)* About eleven in the morning was obſerved an oſcillation in the waters, from north to ſouth.

BOURDEAUX, capital of *Guienne*. A ſhock, or rather a repetition of ſhocks, which laſted ſome minutes.

CAEN, *Normandy*. A great agitation of the *Orne*.

CHARANTE river, *Angoumois*. A commotion in its waters. See *Angouleſme*.

GAINNEVILLE, *Normandy*. (three leagues from *Havre)* A ſenſible oſcillation of the water.

GARONNE, river, *Guienne*. (near *Bourdeaux)* A great agitation of its waters.

HAVRE DE GRACE, *Normandy*. About eleven in the morning, the veſſels in this port were ſtrangely toſſed.

LYONS, capital of *Lyonois*. Divers ſhocks felt here, and in the neighbourhood.

ORNE river, *Normandy*, agitated. See *Caen*, and *Ouilly*.

OUILLY, bridge, *Normandy*. (near *Harcourt)* The waters of the *Orne* much agitated, as alſo thoſe of a lake in this neighbourhood.

IN

In GERMANY.

BRANSTADT, *Holſtein*. The waters were agitated, and the chandeliers in churches were ſeen to vibrate.

EIDER, river, *Holſtein*. An extraordinary commotion of the waters.

ELBE, river. The agitation of the water was ſenſibly perceived through its whole courſe.

EMSHORN, *Holſtein*. Chandeliers vibrated, and waters were diſturbed.

GLUCKSTADT, *Holſtein*. An agitation of the waters which laſted ſeveral minutes.

HAMBURGH. The *Elbe* ſtrongly agitated.

ITZEHOA, *Holſtein*. The waters of the *Stohr* roſe and fell there, and a large float of timber was thrown ſeveral feet on the bank.

OWE river, *Holſtein*. See *Uterſen*.

KELLINGHAUSEN, *Holſtein*. The ſame phenomena as at *Branſtadt* and *Emſhorn*.

LIBBESC lake, *Brandenburg*. The water ebbed and flowed ſix times in half an hour, with a moſt dreadful noiſe, the weather being perfectly calm.

LUBEC, *Holſtein*. Between eleven and twelve, when the wind was at eaſt, and the air quite calm, an extraordinary agitation of the waters was obſerved, particularly in the *Trave*, which roſe four or five feet perpendicular, as it were all at once, by which motion a merchant ſhip ſnapped her cables, and great damage was done to other veſſels. The agitation laſted about nine minutes.

MELDORF, *Holſtein.* The like phænomena as at *Emſhorn* and *Kellengheuſen.*

MUHLGAST lake, *Brandenburg.* The like commotion of the waters as at *Libbeſc* lake.

NETZO lake, *Brandenburg.* The like commotion as at *Libbeſc* and *Muhlgaſt* lakes; but here the waters had an inſupportable ſtench.

RENDSBURG, *Holſtein.* The congregation at divine ſervice in the new church there, obſerved three large chandeliers ſuſpended from the roof, to vibrate very much: Theſe weighed twenty hundred each: A leſſer one over the baptiſmal font was not ſo much affected.

RODDELIN lake, *Brandenburgh*, the like diſturbance of the waters, as *Libbeſc* and *Muhlgaſt* lakes.

SAXONY. Shocks felt in ſeveral of its mines.

STEINBURGH fort, *Holſtein.* In great danger from the violent agitation of the waters which ſurround it.

STOHR or STOUHR river, *Holſtein.* Agitation of its waters. See *Itzehoa.*

STRASBURG, *Alſace.* A ſhock was felt.

STUTGARD, *Wirtemberg.* A ſhock was felt.

TEMPLIN lake, *Brandenburg.* The like phænomena as at *Libbeſc*, *Muhlgaſt*, and *Roddelin* lakes.

TRAVE river, *Holſtein.* Vaſt diſturbance of its waters. See *Lubec.*

UTERSEN, *Holſtein.* A great perturbation in the waters of the *Owe.*

WESER river. Agitations through its whole courſe.

IN HOLLAND.

ALPHEN. (on the *Rhine*, between *Leyden*, and *Woerden)* In the afternoon, the waters were agitated to ſuch a violent degree, that buoys were broken from their chains, large veſſels ſnapped their cables, ſmaller ones were thrown out of the water upon the land, and others lying on land were ſet afloat.

AMSTERDAM. About eleven in the forenoon, the air being perfectly calm, the waters were ſuddenly agitated in their canals, ſeveral boats broke looſe, chandeliers were obſerved to vibrate in the churches, and the mercury which ſtood pretty high in the barometers deſcended almoſt an inch, as it were at once; but no houſe or other building at land was the leaſt ſenſibly ſhaken.

BOIS LE-DUC. Much the ſame motion of the waters as at *Amſterdam*.

BOSHOOP. Nearly the like phænomena as at *Alphen*.

GOUDA (at the confluence of the rivers *Gouw* and *Iſſel)* Much the ſame as at *Amſterdam*.

HARLEM (on the river *Sparen*, a league from the ſea) In the forenoon, for near four minutes together, not only the water in the rivers, canals, &c. but alſo all manner of fluids in ſmaller quantities, as in coolers, tubs, backs, &c. were aſtoniſhingly agitated, and daſhed over the ſides, notwithſtanding no motion was perceptible in the containing veſſels. In ſuch ſmall quantities alſo, the ſurface of the fluid had apparently a direct aſcent, prior to its turbulent motion, and in many places, even the rivers and canals roſe twelve inches perpendicular. In *Harlem* meer the courſe of a veſſel, on full ſail, was ſuddenly ſuſpended, and the rudder unhung.

HAGUE. At eleven in the morning, in abſolutely calm weather, there was obſerved of a ſudden a ſlight motion in the water. A tallow-chandler here heard with ſurprize the claſhing noiſe made by the candles which hung up in his ſhop; but no motion at all was perceived under foot. In a canal between *Delft* and the *Hague*, the riſe was meaſured to be one foot perpendicular.

HERTOGENBOSCH. See *Bois-le-duc*.

LEERDAM. The like as at *Amſterdam*.

LEYDEN. Between half an hour after ten and eleven in the morning, in ſome of the canals of this city, the water roſe ſuddenly on the quay, ſituated on the ſouth. It returned afterwards to its bed, and made ſeveral very ſenſible undulations, ſo that the boats were ſtrongly agitated: the ſame motion was perceived here in the water of the backs of two brew-houſes.

ROTTERDAM. Beſides the like phænomena that were obſerved at *Alphen*, the chandeliers of the *Roman Catholick* church here, which hung from long iron rods, made ſeveral oſcillations.

UTRECHT. The like as at *Alphen*.

WOUBROGGE. The like as at *Alphen*.

In ITALY.

CORSICA iſland. The ſea violently agitated all round it, and moſt of the rivers in the iſland overtopped their banks, and drowned much land. In ſome places a motion of the ground was alſo felt.

LEMAN lake. The waters retired for ſome moments at the end of it.

LODI.. (in the *Milaneſe)* A ſenſible ſhock.

MILAN city. A motion of the earth felt ſeveral times very ſenſibly.

PIZZIGHITONE (in the *Milaneſe)* Shocks felt.

TURIN, *Savoy*. A violent ſhock.

In NORWAY.

VIOLENT agitations of ſeveral rivers and lakes.

In PORTUGAL, and ALGARVE.

THESE kingdoms almoſt univerſally affected, particularly,

BRAGANZA. Much ſhocked and damaged.

CASCAES. (at the mouth of the *Tagus)* Suffered greatly.

COIMBRA. (on the river *Mondego*) About ten in the morning, the ſhocks ſo violent, that the fine building belonging to the *Jesuits*, which consisted of sixteen separate apartments, was almost entirely destroyed, together with the cathedral, and the church of the Holy Cross.

Colares. (about twenty miles from *Lisbon*, behind the rock, about two miles from the sea) The thirty-first of *October* the weather was clear, and uncommonly warm for the season; the wind north, from which quarter, about four o'clock in the afternoon, there arose a fog, which came from the sea, and covered the vallies, a thing rare at this season of the year. Soon after, the wind changing to the east, the fog returned to the sea, collecting it self, and becoming exceeding thick. As the fog retired, the sea rose with a prodigious roaring.

The first of *November*, the day broke with a serene sky, the wind continuing at east: But about nine o'clock, the sun began to grow dim, and about half an hour after was heard a rumbling noise, like that of chariots, which increased to such a degree, as to equal that of the loudest cannon; and immediately a skock of an earthquake was felt, which was succeeded by a second and a third; and several light flames of fire issued from the mountains, resembling the kindling of charcoals. In these three shocks the wall of the building moved from east to west. In another situation from whence the sea-coast could be discovered, there issued from one of the hills, called the *Fojo*, near the beach of *Adraga*, a great quantity of smoke, very thick, but not very black, which still increased with the fourth shock, and

after

after continued to iſſue in a greater or leſs degree. Juſt as the ſubterraneous rumblings were heard, it was obſerved to burſt forth at the *Fojo*; and the quantity of the ſmoke was always proportioned to the noiſe. The place from whence the ſmoke was ſeen to ariſe, was viſited, but it could not be diſcovered from whence it could have iſſued, nor could any ſigns of fire be found near the place: From whence the curious obſerver infers, either that the ſmoke exhaled from ſome eruption or volcano in the ſea, which the waters ſoon covered, or that, if it iſſued from ſome chaſm in the land, it cloſed afterwards. He rather inclines to the former opinion, becauſe it is natural, that the water ſhould retire from the place of the eruption. Beſides, the ſea having riſen in ſome places, it is probable that it fell in others; and indeed it has viſibly retired there, for you may walk on the dry ſhore now, where before you could not wade. And the ſecond conjecture may be true, as ſome chaſms on the dry land are now almoſt cloſed up, and others intirely ſo. In the afternoon preceding the firſt of *November*, the water of a fountain was greatly decreaſed: On the morning of the firſt of *November*, it ran very muddy, and after the earthquake it returned to its uſual ſtate, both in quantity and clearneſs. In ſome places where there was no water, ſprings burſt forth, which continued to run. On the hills numbers of rocks were ſplit, and there were ſeveral rents in the ground, but none conſiderable: On the coaſt pieces of rock fell, ſome of them very large.

DOURO river, ſwelled and overflowed its banks.

ELVAS.

ELVAS. (on the river *Guadiana)* Very much ſhaken and damaged.

FARO. (a ſea-port) A very ſevere ſhock, which overthrew a great number of houſes, and almoſt buried the town in its ruins.

GUADIANA river. Moſt violently agitated. See *Elvas*.

GUIMARANES. (between the *Douro* and the *Minho*) Much ſhaken.

LAGOS. (a ſea-port) Severely ſhaken, and left uninhabitable.

LAMEGO. (near the *Douro)* Suffered much in the ſame manner as *Coimbra* and *Elvas*.

LISBON. *(a)* There was a ſenſible trembling of the earth in 1750, after which it was exceſſively dry for four years together, inſomuch that ſome ſprings, formerly very plentiful of water, were dried, and totally loſt; at the ſame time the predominant winds were eaſt and north-eaſt, accompanied

(a) This city ſuffered greatly by an earthquake in 1531, thus deſcribed by *Paulus Jovius*. hiſt. l. 29. fol. 180.

" In the following month of *January*, a like diſaſter befel " the *Portugueze*, from a ſudden expanſion of wind in the " bowels of the earth, which had well nigh proved fatal to " the city of *Liſbon*; nor did *Azumar*, *Santarein* and *Almerin* " fare much better, for a vaſt number of public edifices and " houſes were ſhaken to pieces and overthrown, and multi- " tudes of the inhabitants buried in the ruins. At the ſame " time there was a horrid ſwell of the ſea, and ſeveral ſhips " were ſucked into the abyſs: The waters of the *Tagus* were " driven on its banks, and the bottom left dry in the middle, " to the unſpeakable amazement of the beholders. The con- " tinual workings of the earth drove almoſt all the inhabitants " of the kingdom out of their houſes, into the open fields, " where, after the example of the royal family, they lived in " tents, not without frequent apprehenſions of being ſwal- " lowed up by the gaping earth.

panied with various, though very ſmall tremors of the earth. The year 1755, proved very wet and rainy, the ſummer cooler than uſual, and for forty days before the great earthquake, clear weather, yet not remarkably ſo. The thirty-firſt of *October*, the atmoſphere, and light of the ſun had the appearance of clouds, with a notable obfuſcation. The firſt of *November*, early in the morning, a thick fog aroſe, which was ſoon diſſipated by the heat of the ſun, no wind ſtirring, the ſea calm, and the weather as warm as in *England* in *June* or *July*. At thirty-five minutes after nine, without the leaſt warning, except a rumbling noiſe, not unlike the artificial thunder at our theatres, immediately preceding, a moſt dreadful earthquake ſhook by ſhort, but quick vibrations, the foundations of all *Liſbon*, ſo that many of the talleſt edifices fell that inſtant : Then, with a ſcarcely perceptible pauſe, the nature of the motion changed, and every building was toſſed like a waggon driven violently over rough ſtones, which laid in ruins almoſt every houſe, church, convent and publick building, with an incredible ſlaughter of the people. It continued in all about ſix minutes. At the moment of the beginning, ſome perſons on the river, near a mile from the city, heard their boat make a noiſe as it run aground or landing, though then in deep water, and ſaw at the ſame time the houſes falling on both ſides the river. Four or five minutes after, the boat made the like noiſe, which was another ſhock, which brought down more houſes. The bed of the *Tagus* was in many places raiſed to its ſurface.

Ships

Ships were drove from their anchors, and joſtled together with great violence; nor did the maſters know if they were afloat or aground. The large new quay, called *Cays Depreda*, was overturned, with many hundreds of people on it, and ſunk to an unfathómable depth in the water, not ſo much as one body afterwards appearing. The bar was ſeen dry from ſhore to ſhore; then ſuddenly the ſea, like a mountain, came rolling in, and about *Belem* caſtle the water roſe fifty feet almoſt in an inſtant, and had it not been for the great bay oppoſite to the city, which received and ſpread the great flux, the low part of it muſt have been under water. As it was, it came up to the houſes, and drove the inhabitants to the hills. About noon, there was another ſhock, when the walls of ſeveral houſes which were yet ſtanding, were ſeen to open from top to bottom more than a quarter of a yard, but cloſed again ſo exactly as to leave ſcarce any mark of the injury.

This earthquake came on three days before the new moon, when three quarters of the tide had run up. The direction of its progreſs ſeems to have been from north to ſouth nearly, for the people on the river, ſouth of the town, obſerved the remoteſt buildings to fall firſt, and the ſweep to be continued down to the waters ſide. Few days paſſed without ſome ſhock for the ſpace of an enſuing year. *October* the tenth, 1756, at eleven at night, there was one which threw down the greateſt part of an hotel, in the pariſh of St. *Andrew*: And *November* the firſt, 1756, being the anniverſary of the fatal tragedy of this unhappy city,

city, another ſhock gave the inhabitants ſo terrible a freſh alarm, that they were preparing for their flight into the country; but were prevented by ſeveral regiments of horſe placed all round by the king's orders.

MONTE	ARRABIDA. ESTRELLA. JULIO. MARVAN. CINTRA.	Theſe, being ſome of the largeſt mountains in *Portugal*, were impetuouſly ſhaken, as it were from their very founda-

tions, and moſt of them opened at their ſummits, ſplit and rent in a wonderful manner, and huge maſſes of them were thrown down into the ſubjacent vallies.

OPORTO. (near the mouth of the *Douro)* At about forty minutes paſt nine in the morning, the ſky very ſerene, was heard a dreadful hollow noiſe like thunder or the rattling of coaches over rugged ſtones at a diſtance; and almoſt at the ſame inſtant was felt a ſevere ſhock of an earthquake, which laſted ſix or ſeven minutes, during which ſpace every thing ſhook and rattled. It rent ſeveral churches, and tumbled down one of the turrets of that of the *Congregadoes*. In the ſtreets the earth was ſeen to heave under people's feet, as if in labour. The river was alſo amazingly affected; for in the ſpace of a minute or two, it roſe and fell five or ſix feet, and continued ſo to do for four hours. It ran up at firſt with ſo much violence, that it broke a ſhip's hawſer. Two of the *Brazil* fleet were going out, and had got to the bar, but the ſea impetuouſly forced them back again into the harbour, drove them foul of one

 another,

another, and they narrowly escaped being lost. The river was observed to burst open in some parts, and discharge vast quantities of air; and the agitation was so great in the sea, about a league beyond the bar, that 'tis imagined the air got vent there too. Two other shocks followed this first the same day, but they were short, and much slighter.

PEDRA DE ALVIDAR. (a rock near the hill *Fojo*; see *Colares)* A kind of parapet was broken off from it, which issued from its foundation in the sea.

SANTAREIN. (on the *Tagus*) Suffered much.

SARITHOES and BITURECRAS. Two rocks in the sea near the mouth of the *Tagus*, one of them was broken off at the summit, the other all to pieces.

SETUVAL, SAINT UBAL, or SAINT UBES. (a sea port twenty miles south of *Lisbon*) No traces left of this place, the repeated shocks, and vast surf of the sea having concurred to swallow it up, people and all; which it could the less withstand, as it stood at the head of a little gulph formed by the tide at the mouth of the *Zadaon*. Huge pieces of rock were detached at the same time from the promontory on the west of the town, which consists of a chain of mountains containing fine jasper of different colours.

SILVAS. (four leagues from *Lagos*) Almost entirely destroyed.

TAGUS river, swelled and agitated throughout its whole course, for the space of 300 miles.

VARGE. (on the river *Macaas*) At the time of the earthquake many springs of water burst forth, and some spouted to the height of eighteen or twenty feet, throwing up ſand of various colours, which remained on the ground.

VIANA, (a ſea-port at the mouth of the *Lima*) very much damaged.

VILLA NOVA. (two leagues from *Lagos*) Met with almoſt the ſame fate as *Faro*.

VILLA REAL (four leagues to the north of *Lamego*) much ſhattered.

ZIZAMBRE. A mountainous point ſeven or eight leagues from *Setuval*; which cleft aſunder and threw off ſeveral vaſt maſſes of rock.

IN SPAIN.

FELT all over it, except in *Catalonia*, *Aragon* and *Valencia*, more particularly at

ALGAZAIST. (at the *Streight*'s mouth) Several walls fell down, and great part of the town was overflowed.

ANTEQUERA. (on a mountain in *Granada*, five leagues north of *Malaga)* Greatly damaged.

ARCOS. (on the *Guadalete*) Much ſhattered.

AYAMONTE. (near where the *Guadiana* falls into the bay of *Cadiz*) A little before ten o'clock, immediately upon a hollow ruſhing noiſe being heard, a terrible earthquake was felt, which during fourteen or fifteen minutes, damaged almoſt all the buildings, throwing ſome down, and leaving others irreparably ſhattered. In little more than half an hour after, the ſea and river, with all their canals, overflowed their bounds with great violence, laying under water all the coaſts of of the iſlands adjacent to the city and its neighbourhood, flowing into the very ſtreets. The water roſe three times, after it had as many times ſubſided. One of the ſwells was at the time of ebb, and viſibly with leſs violence. The water came on in vaſt black mountains, white with foam at the top, and demoliſhed more than half of the tower at the bar called *de Canala*. The earth was obſerved to open in ſeveral parts, and from the apertures flowed large quantities of water, eſpecially in the maritime places. In the adjacent ſtrands the damage was much greater, as the ſea ſwallowed up all the huts built there, deſtroying goods and treaſure beyond redemption; for all that was overflowed ſunk, and the beach became a ſea, without the leaſt ſign of what it was before. Many perſons periſhed, for although they got aboard ſome veſſels, yet part of theſe foundered, and others being forced out to ſea, the unhappy paſſengers were ſo terrified that they threw themſelves over-board. The day was ſerene, and not a breath of wind ſtirring.

BILBOA. (on the *Nervio*, two leagues from the ocean) A ſhock and commotion of the waters.

CADIZ. (at the north-weſt end of the iſland of *Leon*, oppoſite to *Port Saint Mary*) Some minutes after nine in the morning, the whole town was ſhook with a violent earthquake, which laſted about five minutes. The water in the ciſterns under ground waſhed backwards and forwards, ſo as to make a great froth upon it. At ten minutes after eleven, a wave was ſeen coming from ſea, eight miles off, at leaſt ſixty feet higher than common.

common. It dashed against the west part of the town, which is very rocky, and the rocks abated a great deal of its force: At last it came upon the walls, beat in the breast-work, and carried pieces of eight or ten ton weight, forty or fifty yards from the wall, bore away the sand and walls, but left the houses standing, being exceeding strong built. The governor ordered the gates to be shut, that people might not go out of the town, as the land was lower than the town, by which he saved the lives of thousands. When the wave was gone, some parts that are deep at low water, were left quite dry, for the water returned with the same violence it came. At thirty minutes after eleven came a second tide; and these two were followed by four others of the same kind, at eleven o'clock fifty minutes; twelve o'clock thirty minutes; one o'clock ten minutes; and one o'clock fifty minutes. The tides continued, with some intervals, till the evening, but lessening. Every thing was washed off the mole. There was a strong causey on a very narrow neck of land that goes from the town to the isle of *Leon*, open to the sea on one side, and to the bay on the other, which was washed away, and scarce any mark of it left. About forty or fifty persons, and many cattle that were on it, were all drowned. The ships were exposed to imminent danger; the greatest part of them were driven afloat, but most of them fortunately were saved, some by veering their cables, others by securing themselves by new anchors; so that only one *Swedish* ship, and some boats were

 oft,

loſt. The whole day was as clear and ſerene as at midſummer, without a breath of wind.

CHICLAN (in the iſle of *Cadiz*) ſhocked and overflowed.

CONIL (a ſmall port five leagues ſouth of *Cadiz*) ruined.

CORDOUA (on the *Guadalquivir*) greatly damaged.

ESCURIAL. (the moſt magnificent of the king's palaces, ſeven leagues north-weſt of *Madrid*) Moſt terrible ſhocks, felt by all the royal family, which occaſioned their immediate removal.

ESTAPONA (on the *Mediterranean* ſea-coaſt, between *Marbella* and *Gibraltar*) the earthquake greatly damaged the church.

GIBRALTAR. (in the *Straits* mouth) About ten minutes after ten, a tremulous motion of the earth was plainly perceived, which laſted about half a minute, then a violent ſhock, after that a trembling for five or ſix ſeconds, then another ſhock not ſo violent as the firſt, which went off gradually as it began. It laſted, in the whole, about two minutes. The guns on the battery were ſeen ſome to riſe, others to ſink, the earth having an undulating motion. Moſt people were ſeized with giddineſs and ſickneſs, and ſome fell down, others were ſtupified, though many that were walking or riding felt no motion, but were ſick. The ſea roſe ſix feet every fifteen minutes, and fell ſo low that boats and all the ſmall craft near the ſhore were left aground, as were numbers of ſmall fiſh. Ships out in the bay thought they had ſtruck upon rocks. This flux and reflux laſted

laſted till next morning, having decreaſed gradually from two in the afternoon. The day was clear, and but little wind at ſouth-weſt. *Fahrenheit*'s thermometer was at ſixty-two, and no alteration was obſerved.

GRANADA (on the river *Xènil)* damaged conſiderably.

MADRID. (capital of all *Spain*, on the *Manzanares)* Five minutes after ten in the morning, a great earthquake was very ſenſibly felt, which laſted about ſix minutes. Every body at firſt thought they were ſeized with a ſwimming in their heads; and afterwards that the houſes in which they were, were falling. The ſame happened in the churches, ſo that people trod one another under foot in getting out; and thoſe who obſerved it in the towers, were very much frightened, thinking that they were tumbling to the ground. It was not felt in coaches, nor, but very little, by thoſe who walked on foot. No remarkable accident happened, excepting that two lads were killed by the falling of a ſtone croſs from the porch of a church belonging to a monaſtry. St. *Andrews* church was ſo much ſhaken, that ſeveral apertures remain in the roof and walls; the upper part of the porch of the pariſh church of St. *Lewis* was ſplit; and thoſe of St. *Philip*, St. *Thomas*, *Portaceli*, and the towers of St. *Trinity* and St. *Millan*, were forced to be examined by ſkilful workmen.

MALAGA (a ſea-port on the *Mediterranean)* felt a violent ſhock; the bells rung in the ſteeples; the water overflowed in a well, and as suddenly retired again.

MEDINA SIDONIA (nine leagues from *Cadiz*) severely shocked.

PORT-REAL. (near *Cadiz*) Much shocked and inundated.

PORT SAINT MARY. (at the mouth of the *Guadalete*) The sea rose and subsided eight several times.

PURVELO. (near *Saint Lucar*) Its steeple and several houses shaken down.

SALAMANCA. (on the *Tormes*, thirty-three leagues north-west of *Madrid*) Shocks felt, and the waters agitated.

SANT LUCAR. (at the mouth of the *Guadalquivir*) Violent shocks, and the sea broke in and did great mischief.

SANT ROQUE. A smart shock which tossed persons out of their seats, and rent an arch of the church.

SEGOVIA. (on the *Elrena*, ten leagues north of *Madrid*) A great commotion of the waters.

SEVILLE. (on the *Guadalquivir*, sixteen leagues above the mouth of it) The earthquake shook down several houses, and greatly damaged some churches, especially the cathedral, the finest in the kingdom, whose famous tower, called *la Giralda*, opened in the four sides, and a great many large stones falling down, killed several persons. The waters were so greatly agitated, that all the vessels in the river were driven ashore.

VALENCIA. (on the *Savar*) Very terrible agitations of the water.

TOLEDO (on the *Tagus*, fourteen leagues ſouth of *Madrid*) the river roſe ten feet.

XERES (on the *Guadalate*, ſix leagues north of *Cadiz*) much ſhaken and damaged.

IN SWEDEN.

THE earthquake was felt in ſeveral provinces, and all the rivers and lakes were ſtrongly agitated, eſpecially in *Dalecarlia*.

DALA river. Its waters overflowed the adjacent fields, and afterwards retired within its bed, with no leſs rapidity. At the ſame time a lake a league diſtant from it, and which had no manner of communication with it, bubbled up with great violence.

FAHLUN. (in *Dalecarlia*) Several ſtrong ſhocks were felt during the time of divine ſervice.

IN SWISSERLAND.

MANY rivers were ſuddenly turned muddy without rain.

NEUFCHATEL. Its lake ſwelled to the height of near two feet above its natural level, for the ſpace of a few hours.

ZURICH. An agitation was perceived in the waters of its lake.

AFRICA.

ALGIERS. Great part of it deſtroyed.

ARZILA. About ten in the morning the ſea came ſuddenly up, and ſeven *Moors*, who were out of the town walls, were drowned; the waters came through one of the city gates very far. It roſe with ſuch impetuoſity, that it lifted up a veſſel in the bay, which, at the waters falling down again, it dropped with ſuch force upon the land, that it was broke to pieces; and a boat was found at the diſtance of two muſket-ſhot within land from the ſea.

FEZ. Vaſt numbers of houſes fell down, and a great multitude of people were buried in the ruins.

MEQUINEZ. Two thirds of the houſes fell down, and alſo the convent of the *Franciſcan Friers*. Many lives were loſt.

MOROCCO. By the falling down of a great number of houſes many people loſt their lives; and about eight leagues from this city, the earth opened, and ſwallowed up a village, with all the inhabitants (who were known by the name of the *Sons of Buſunba*) to the number of about eight or ten thouſand perſons, together with their cattle of all ſorts, as camels, horſes, horned beaſts, &c. and ſoon after the earth cloſed again, in the ſame manner as it was before.

SAFFE. Several houſes fell down, and the ſea came up as far as the great Moſque, which is within

within the city, and at a great diſtance from the ſea.

SALLE. The damage here was very great, near a third part of the houſes having been overthrown. The waters came into the city with great rapidity, and at their falling off great quantities of fiſh were found in the ſtreets, and many perſons were drowned: Two ferry-boats were overſet in the river, and all the people on board were alſo drowned; and a large number of camels that were juſt then going for *Morocco*, were carried away by the waters.

SARJON hills. One of theſe was rent in two; one ſide of which fell upon a large town, where there was the famous ſanctuary of their prophet, called *Mulay Teris*; and the other ſide fell down upon another large town, and both towns and the inhabitants were all buried under the ſaid hill.

SCLOGES. (a place where the *Barbarians* live, not far from *Fez)* A mountain broke open, and a ſtream iſſued out as red as blood.

TANGIER. The earthquake began at ten in the morning, and laſted ten or twelve minutes. The trembling of the houſes, moſques, &c. was great, and a large projecting part of an old building near the city gate, after three ſhocks fell down to the ground. The ſea came up to the very walls, a thing never ſeen before, and went down directly with the ſame rapidity as it roſe, as far as the place where the large veſſels anchor in the bay, leaving upon the mole a great quantity of ſand and fiſh. Theſe commotions of the ſea were repeated eighteen times, and continued till ſix in the

 evening,

evening, though not with ſuch violence as at the firſt time. The fountains were dryed up, ſo that there was no water to be had till night: And as to the ſhore ſide, the waters came up half a league in land.

TETUAN. The earthquake began here at the ſame time as at *Tangier*, but laſted only between ſeven and eight minutes, during which ſpace the ſhock was repeated three different times, with ſuch violence, that it was feared the whole city would fall down: It was likewiſe obſerved, that the waters of the river *Chico*, on the other ſide of the city, and thoſe of a fountain, appeared very red.

IN THE ATLANTIC ISLANDS.

ANTIGUA. About the time of the earthquake at *Liſbon*, there was ſuch a ſea without the bar of this iſland, as had not been known in the memory of man; and after it all the water at the wharfs, which uſed to be ſix feet, was not two inches.

BARBADOES. About two o'clock in the afternoon, the ſea ebbed and flowed in a moſt ſurprizing manner. It ran over the wharfs and the ſtreets into the houſes, and at the old bridge brought up numbers of ſeveral ſorts of fiſh. It continued thus ebbing and flowing till ten at night.

MADEIRA. In the city of *Funchal*, thirty-eight minutes paſt nine in the morning, was perceived a ſhock of an earthquake; the firſt notice whereof was a rumbling noiſe in the air, like that of empty carriages paſſing haſtily over a ſtone pavement. The obſerver felt the floor immediately to move with a tremulous motion, vibrating very quickly: The windows rattled, and the whole houſe ſeemed to ſhake; it laſted more than a minute, during which, the vibrations, though continual, abated and increaſed twice very ſenſibly, in point of force: not unlike an eccho from the diſcharge of a fowling-piece, oppoſite to a range of mountains, whence the ſound has reverberated with reciprocal intenſions and remiſſions. The increaſe, after the firſt remiſſion of the ſhock, was the moſt intenſe: The door of the room vibrating to and fro very remarkably then, which it had not done before; neither did it afterwards in the ſecond increaſe. The noiſe in the air, which had preceded the ſhock, continued to accompany it; and laſted ſome ſeconds after the motion of the earth had entirely ceaſed; dying away like a peal of diſtant thunder rolling through the air. The direction of the ſhock ſeemed to be from eaſt to weſt. At three quarters paſt eleven, the ſea, which was quite calm (it being a fine day and no wind ſtirring) was obſerved to retire ſuddenly ſome paces; then riſing, with a great ſwell, without the leaſt noiſe, and as ſuddenly advancing, overflowed the ſhore, and entered the city. It roſe full fifteen feet perpendicular above high-water mark, although the tide, which ebbs and flows there ſeven feet, was then at half ebb. The water immediately receded again, and, after having fluctuated four or five times between high-water and low-water mark,

mark, the undulations continually decreaſing (not unlike the vibrations of a pendulum) it ſubſided, and the ſea remained calm, as before this phænomenon. The ſeaſon of the year had been more than ordinary dry; the rains, which generally begin to fall the beginning of *October*, not having then ſet in. The weather for ſome weeks preceding the earthquake, had been very fine and clear, but the day previous thereto, (*October*, 31) was very remarkably fair and ſerene, as was the former part of the day on which it happened: But the afternoon was very dull and dark, the ſky being entirely overcaſt with heavy black clouds; the ſubſequent day was very fair. The greateſt height of *Fahrenheit*'s thermometer, the three laſt days of *October*, and the firſt of *November* was 69. *November* the ſecond, it roſe to 71. The barometer had been ſtationary ſeveral days at 29,28 inch. *November* the ſecond, it roſe to 30,1. In the northern part of the iſland the inundation was more violent, the ſea there retiring above one hundred paces at firſt, and ſuddenly returning, overflowed the ſhore, forcing open doors, breaking down the walls of ſeveral magazines and ſtorehouſes, and carrying away in its receſs a conſiderable quantity of grain and ſome hundred pipes of wine. Great quantities of fiſh were left aſhore, and in the ſtreets of the village of *Machico*. All this was the effect of one inundation of the ſea, which never flowed afterwards ſo high as high-water mark; although it continued fluctuating there much longer before it ſubſided, than at *Funchal*, as the fluctuation and ſwell was much greater

ter at *Funchal*, than it had been farther to the weſtward, where, in ſome places, it was hardly, if at all, perceptible.

SAINT MARTINS. The earthquake ſlightly felt.

TERCERA. Some ſhocks felt.

It has been reported that much damage was done in the *Canary* iſlands, but no particulars have as yet come to hand.

AT SEA, and in the OCEAN.

OFF St. *Lucar*. The captain of the *Nancy* felt his ſhip ſo violently ſhaken, that he thought ſhe had ſtruck the ground; but after heaving the lead overboard, found ſhe was in a great depth of water.

Captain *Clark* from *Denia*, in latitude 36°. 24. between nine and ten in the morning, had his ſhip ſhaken and ſtrained as if ſhe had ſtruck on a rock, ſo that the ſeams of the deck opened, and the compaſs was overturned in the benacle.

The maſter of a veſſel bound to the *American* iſlands, being in latitude 25°. N. longitude 40°. and writing in his cabin, heard a violent noiſe, as he imagined, in the ſteerage; and whilſt he was aſking what was the matter, the ſhip was put into a ſtrange agitation, and ſeemed as if ſhe had been ſuddenly jerked up, and ſuſpended by a rope faſtened to the maſt head. He immediately ſtarted up with great terror and aſtoniſhment, and looking out at the cabbin window, plainly diſcovered land

at

at the diſtance of about a mile; upon this he haſtily ordered the lead to be thrown, ſuppoſing the ſhip might have ſtruck; but coming upon deck, the land he had ſeen was no more to be found, and he perceived with great amazement a violent current croſs the ſhip's way to the leeward. In about a minute this current returned with great impetuoſity, and within a league he ſaw three craggy pointed rocks, throwing up water of various colours, reſembling liquid fire. This phænomenon in about two minutes ended in a black cloud, which aſcended very heavily. After it had riſen above the horizon, no rock was to be ſeen; and the agitation of the water ſoon ſubſided, tho' the cloud, ſtill aſcending, was long viſible, the weather being extremely clear.

The captain of a *Dutch* veſſel, which had ſailed from St. *Ubes*, about eight in the morning, being at a quarter after ten, near a league and a half from mount *Sizembre*, which is about ſix or ſeven leagues from St. *Ubes*, felt a violent ſhock in his ſhip, and at the ſame time ſaw that mountain rend, and ſeveral large rocks rowl from it into the ſea, with a vaſt and horrid noiſe. Immediately after, the ſky was covered with a thick fog, occaſioned by the fall of the rocks into the water. The ſhock was repeated at different intervals, till ſun-ſet, at which time he obſerved a thick ſmoke at N. N. E. diſtant ſeven or eight leagues, and ſoon after flames, which continued all night. The light of the ſun, and the diſtance intercepted them from his ſight next morning.

In

In latitude 38°. N. 10°. 47 W. off cape St. *Vincent*, at half an hour paſt nine, a ſhip felt a terrible ſhock which laſted three minutes, and more ſhocks till half an hour paſt eleven, all attended with a growling noiſe. The ſky was ſerene, and the ſea ſmooth: This was out of ſoundings.

Between nine and ten in the morning, forty leagues weſt of the ſame cape, in a calm ſea, another ſhip was ſo violently agitated, that the anchors, which were laſhed, bounced up, and the men were thrown a foot and an half along the deck; and of a ſudden the ſhip ſunk in the water, as low as her main chains. The lead ſhewed a great depth of water, and the line was tinged of a yellow colour, and ſmelt of ſulphur. This ſhock laſted about ten minutes, but they felt ſmaller ones for about twenty-four hours.

Several *Dutch* ſhips off cape St. *Mary*, thought they ſtruck aground, and fired guns of diſtreſs.

Of the extent of this EARTHQUAKE.

WE have ſeen that, beſides a multitude of other places, it was very ſenſible in Europe at *Fahlun* in *Sweden*, in *Africa* at the capital of the empire of *Morocco*, and in *America* at the iſland of *Barbadoes*. Between *Fahlun* and *Barbadoes* are ſeventy degrees of a great circle, nearly; between *Barbadoes* and *Morocco* forty-nine, and between *Morocco* and *Fahlun* thirty-three of the like degrees: Now theſe conſtitute the three ſides of of a ſpherical triangle, to which if a well known theorem be applyed, it will be found, that the effects of the earthquake of the firſt of *November*, one thouſand ſeven hundred and fifty-five, were diſtributed over very nearly four millions of ſquare *Engliſh* miles of the earth's ſurface: A moſt aſtoniſhing ſpace! and greatly ſurpaſſing any thing, of this kind, ever recorded in hiſtory.

The END.

ERRATA.

Page 12. line 22. read, ἐνοσίγαιον. p. 22. l. 10. r. doctrines. p. 23. l. 13. r. meteorologics. p. 25. l. 1. r. ſteam. p. 29. line laſt but one, for in, r. on. p. 64. l. 31. for If, r. Of. p. 67. l. 14. r. *nubigenum*. p. 88. l. laſt but one, r. *Aſbeſtus*. p. 106. l. 25. r. *Varenius*. p. 127. l. 11. r. *Achaia*. p. 250. l. 11. r. the ſuperficial. r. *Camden* throughout.

INDEX

INDEX.

 most

2 *Norcia*,

Vernatti

CONJECTURES
CONCERNING THE
CAUSE,
AND
OBSERVATIONS
UPON THE
PHÆNOMENA,
OF
EARTHQUAKES;

Particularly of

That great Earthquake of the firſt of November 1755, which proved ſo fatal to the City of Liſbon, and whoſe Effects were felt as far as Africa, and more or leſs throughout almoſt all Europe.

By the Reverend JOHN MICHELL, M. A.
Fellow of Queen's-College, Cambridge.

Read at ſeveral Meetings of the ROYAL SOCIETY.

LONDON:
Printed in the Year M.DCC.LX.

CONJECTURES concerning the CAUSE,

AND

OBSERVATIONS upon the PHÆNOMENA,

OF

EARTHQUAKES.

INTRODUCTION.

Read Feb. 28. March 6. 13. 20. 27. 1760.

ART. I. IT has been the general opinion of philosophers, that earthquakes owe their origin to some sudden explosion in the internal parts of the earth. This opinion is very agreeable to the phænomena, which seem plainly to point out something of that kind. The conjectures, however, concerning the cause of such an explosion, have not been yet, I think, sufficiently supported by facts; nor have the more particular effects, which will arise from it, been traced out; and the connexion of them with the phænomena explained. To do this, is the intent of the following pages; and this we are now the better enabled to do, as the late dreadful earthquake of the

1st

1ſt of November 1755 ſupplies us with more * facts, and thoſe better related, than any other earthquake of which we have an account.

2. That theſe concuſſions ſhould owe their origin to ſomething in the air, as it has ſometimes been imagined, ſeems very ill to correſpond with the phænomena. This, I apprehend, will ſufficiently appear, as thoſe phænomena are hereafter recounted; nor does there appear to be any ſuch certain and regular connexion between earthquakes and the ſtate of the air, when they happen, as is ſuppoſed by thoſe who hold this opinion. It is ſaid, for inſtance, that earthquakes always happen in calm ſtill weather: but that this is not always ſo, may be ſeen in an account of the † earthquakes in Sicily of 1693, where we are told, "the ſouth winds have blown very much, which ſtill "have been impetuous in the moſt ſenſible earth-"quakes, and the like has happened at other times."

3. Other examples to the ſame purpoſe we have in an account of the earthquakes that happened in New England in 1727 and 1728; the author of

* Many of theſe facts are collected together in the 49th volume of the Philoſophical Tranſactions. The ſame are alſo to be found, with ſome additional ones, in "The Hiſtory and Philoſophy of "Earthquakes," (a work well worth the peruſal of thoſe, who are deſirous of being acquainted with this ſubject). The author of it has given us, beſides the aforeſaid facts, a very judicious abridgment of ten of the moſt conſiderable writers upon the ſubject. I have taken the greateſt part of my authorities either from this author, or the Philoſophical Tranſactions, that thoſe who would wiſh to examine them, may have an opportunity of doing it the more eaſily; ſome things only, which were not to be met with in theſe, and which yet were neceſſary to my purpoſe, I have been obliged to ſeek for elſewhere.

† See Phil. Tranſ. N° 207. or vol. ii. p. 408. Lowthorp's Abr.

which

which ſays, that he could neither obſerve any connexion between the weather and the earthquakes, nor any prognoſtic of them; for that they happened alike in all kinds of weather, at all times of the tides, and at all times of the moon *.

4. If, however, it ſhould ſtill be ſuppoſed, notwithſtanding theſe inſtances to the contrary, that there is ſome general connexion between earthquakes and the weather, at the time when they happen, yet, ſurely, it is far more probable, that the air ſhould be affected by the cauſes of earthquakes, than that the earth ſhould be affected in ſo extraordinary a manner, and to ſo great a depth; and that this,

* See Philoſ. Tranſ. N° 409. or vol. vi. part ii. p. 202. Eames's Abridgment.—To theſe authorities, we may add the opinion of Monſ. Bertrand, who expreſſes himſelf, upon this occaſion, in the following manner. "Ariſtotle, Pliny, and "Seneca, tell us, that earthquakes are preceded by a calm and "ſerene air. This is, indeed, often the caſe, but not always. "I don't know, upon an examination of the whole, if there are "not as many exceptions to this rule, as examples that confirm "it. Some authors again have thought, that they might look on "a dark ſky, lightenings, and ſudden ſtorms, as the forerunners "of earthquakes" Then relating ſome inſtances of ſhocks that happened in calm and ſerene weather, he adds, "On the other "hand, it appears, from the examples, which we have before "related, that many earthquakes have happened at the time of "great rains, violent winds, and with a cloudy ſky; ſo that one "cannot find any certain prognoſtic of them in the ſtate of the "atmoſphere." See *Memoires Hiſtoriques et Phyſiques ſur les tremblemens de Terre, par Monſ. Bertrand, a la Haye* 1757. This author, in theſe ſenſible memoirs, has obliged the public with a circumſtantial account of all the facts he could collect, relating to the earthquakes of Switzerland, or thoſe of other places, that ſeemed to be connected with them. The whole ſeems to be done with care and fidelity, and without the leaſt attachment to any particular ſyſtem.

and

and all the other circumſtances attending theſe motions, ſhould be owing to ſome cauſe reſiding in the air.

5. Let us then, rejecting this hypotheſis, ſuppoſe, that earthquakes have their origin under ground, and we need not go far in ſearch of a cauſe, whoſe real exiſtence in nature we have certain evidence of, and which is capable of producing all the appearances of theſe extraordinary motions. The cauſe I mean is ſubterraneous fires. Theſe fires, if a large quantity of water ſhould be let out upon them ſuddenly, may produce a vapour, whoſe quantity and elaſtic force may be fully ſufficient for that purpoſe. The principal facts, from which I would prove, that theſe fires are the real cauſe of earthquakes, are as follow.

SECTION I.

6. *Firſt*, The ſame places are ſubject to returns of earthquakes, not only at ſmall intervals for ſome time after any conſiderable one has happened, but alſo at greater intervals of ſome ages.

7. Both theſe facts ſufficiently appear, from the accounts we have of earthquakes. The tremblings and ſhocks of the earth at * Jamaica in 1692, at * Sicily in 1693, and at * Liſbon in 1755, were repeated ſometimes at larger, and ſometimes at ſmaller intervals, for ſeveral months. The ſame thing has been obſerved in all other very violent earthquakes. At † Lima, from the 28th October 1746, to the

* See the accounts of theſe in the Philoſ. Tranſ.

† See Antonio d'Ulloa's Voyage to Peru, part ii. book i. ch. 7.

24th

24th February 1747 (the time when the account of them was ſent from thence), there had been numbered no leſs than 451 ſhocks, many of them little inferior to the firſt great one, which deſtroyed that city.

8. The returns of earthquakes alſo, in the ſame places, at larger diſtances of time, are confirmed by all hiſtory. Conſtantinople, and many parts of Aſia Minor, have ſuffered by them, in many different ages: Sicily has been ſubject to them, as far back as the remains even of fabulous hiſtory can inform us of: Liſbon did not feel the effects of them for the firſt time in 1755: Jamaica has frequently been troubled with them, ſince the Engliſh firſt ſettled there; and the Spaniards, who were there before, uſed to build their houſes of wood, and only one ſtory high, for fear of them: * Lima, Callao, and the parts adjacent, were almoſt totally deſtroyed by them twice, within the compaſs of about ſixty years, ſcarce any building being left ſtanding, and the latter being both times overflowed by the ſea: nor were theſe the only inſtances of the like kind, which have happened there; for, from the year 1582 to 1746, they have had no leſs than ſixteen very violent earthquakes, beſides an infinity of leſs conſiderable ones; and the Spaniards, at their firſt ſettling there, were told by the old inhabitants, when they ſaw them building high houſes, that they were building their own ſepulchres †.

9. *Secondly,*

* See the place above-quoted.

† What is here ſaid, is taken from d'Ulloa's Voyage to Peru, the Hiſtory and Philoſophy of Earthquakes, the Philoſ. Tranſ. *&c.*

whers

9. *Secondly*, Thoſe places that are in the neighbourhood of burning mountains, are always ſubject to frequent earthquakes; and the eruptions of thoſe mountains, when violent, are generally attended with them.

10. Aſia Minor and Conſtantinople may be looked upon as in the neighbourhood of Santerini. The countries alſo about * Ætna, Veſuvius, mount Hæcla, &c. afford us ſufficient proofs to the ſame purpoſe. But, of all the places in the known world, I ſuppoſe, no countries are ſo ſubject to earthquakes, as † Peru, Chili, and all the weſtern parts of South America; nor is there any country in the known world ſo full of volcanos: for, throughout all that long range of mountains, known by the name of the Andes, from 45 degrees ſouth latitude, to ſeveral degrees north of the line, as alſo throughout all Mexico, being about 5000 miles in extent, there is a continued chain of them ‡.

11. *Thirdly*, The motion of the earth in earthquakes is partly tremulous, and partly propagated by waves, which ſucceed one another ſometimes at larger and ſometimes at ſmaller

---

where many more examples, to the ſame purpoſe, are to be met with. See alſo *Memoires ſur les tremblemens de Terre*; in which are mentioned above 130 repetitions of earthquakes, that have happened, within the compaſs of 960 years, in Switzerland.

* See many inſtances of this in vol. ii. of Lowthorp's Abr. of the Philoſ. Tranſ.

† Monſ. Bouguer ſays, that ſcarce a week paſſes without earthquakes in ſome part of Peru. See Hiſt. of Earthq. p. 205.

‡ See the Maps of theſe countries, Condamine's Voyage down the Maranon, Acoſta's Nat. Hiſt. of the Indies, &c.

diſtances; and this latter motion is generally propagated much farther than the former.

12. The former part of this propoſition wants no confirmation: for the proof of the latter, *viz.* the wave-like motion of the earth, we may appeal to many accounts of earthquakes: it was very remarkable in the two, which happened at Jamaica in * 1687-8 and * 1692. In an account of the former, it is ſaid, that a gentleman there ſaw the ground riſe like the ſea in a wave, as the earthquake paſſed along, and that he could diſtinguiſh the effects of it, to ſome miles diſtance, by the motion of the tops of the trees on the hills. Again, in an account of the latter, it is ſaid, " the ground heaved and ſwelled, " like a rolling ſwelling ſea," inſomuch, that people could hardly ſtand upon their legs by reaſon of it.

13. The ſame has been obſerved in the earthquakes of † New England, where it has been very remarkable. A gentleman giving an account of one, that happened there the 18th November 1755, ſays, the earth roſe in a wave, which made the tops of the trees vibrate ten feet, and that he was forced to ſupport himſelf, to avoid falling, whilſt it was paſſing.

14. The ſame alſo was obſerved at ‡ Liſbon, in the earthquake of the 1ſt November 1755, as may be

---

* See Phil, Tranſ. N° 209. or vol. ii. Lowthorp's Abridgment, p. 410.

† See Philoſ. Tranſ. vol. l. p. 1, &c.

‡ See the accounts collected together, in the 49th volume of the Philoſ. Tranſ. or in Hiſt. and Philoſ. of Earthq. and particularly p. 315. where it is ſaid, " A moſt dreadful earthquake ſhook by

 " ſhort,

be plainly collected from many of the accounts that have been publiſhed concerning it, ſome of which affirm it expreſly: and this wave-like motion was propagated to far greater diſtances than the other tremulous one, being perceived by the motion of waters, and the hanging branches in churches, through all Germany, amongſt the Alps, in Denmark, Sweden, Norway, and all over the Britiſh iſles.

15. *Fourthly*, It is obſerved in places, which are ſubject to frequent earthquakes, that they generally come to one and the ſame place, from the ſame point of the compaſs. I may add alſo, that the velocity, with which they proceed, (as far as one can collect it from the accounts of them) is the ſame; but the velocity of the earthquakes of different countries is very different.

16. Thus all the ſhocks, that ſucceeded the firſt great one at Liſbon in 1755, as well as the firſt itſelf, came from the * north-weſt. This is aſſerted by the perſon, who ſays, he was about writing a hiſtory of the earthquakes there: all the other accounts alſo confirm the ſame thing; for what ſome ſay, that they came from the north, and others, that they came

---

" ſhort, but quick vibrations, the foundations of all Liſbon;
" then, with a ſcarcely perceptible pauſe, the nature of the mo-
" tion changed, and every building was toſſed like a waggon
" driven violently over rough ſtones, which laid in ruins almoſt
" every houſe, church, *&c.*"

For the wave-like motion at Oporto, ſee Phil. Tranſ. vol. xlix. p. 418. for the ſame at Gibraltar, ſee Hiſt. and Philoſ. of Earthq. p. 322.

* See Philoſ. Tranſ. vol. xlix. p. 410.

from

from the weſt, cannot be looked on as any reaſonable objection to this, but rather the contrary. The velocity alſo, with which they were all propagated, was the ſame, being at leaſt equal to that of ſound; for they all followed * immediately after the noiſe that preceded them, or rather the noiſe and the earthquake came together: and this velocity agrees very well with the intervals between the time when the firſt ſhock was felt at Liſbon, and the time when it was felt at other diſtant places, from the compariſon of which, it ſeems to have travelled at the rate of more than † twenty miles *per* minute.

17. An hiſtorical account of the earthquakes, which have happened in ‡ New England, ſays, that, of five conſiderable ones, three are known to have come from the ſame point of the compaſs, *viz.* the north-weſt: it is uncertain from what point the other two came, but it is ſuppoſed that they came from the ſame with the former. The ‖ velocity of theſe has been much leſs than that of the Liſbon earthquakes: this appears from the interval between the preceding noiſe, and the ſhock, as well as from the wave-like motion before-mentioned.

---

* See Philoſ. Tranſ. vol. xlix. p. 414. or Hiſt. and Philoſ. of Earthq. p. 315.

† See Art. 97.

‡ See Philoſ. Tranſ. vol. l. p. 9.

‖ As in ſome earthquakes the velocity, with which they are propagated, is much leſs than in others, it is evident, that they can by no means be owing to any cauſe reſiding in the air: for any ſhock communicated to the air, muſt neceſſarily move with a velocity neither greater nor leſs than that of ſounds; that is, at the rate of about thirteen miles *per* minute.

18. All the greater earthquakes, that have been felt at * Jamaica, ſeem, by the accounts given of them, to have come from the ſea, and, paſſing by Port-Royal, to have gone northwards. The velocity of theſe alſo was far ſhort of the velocity of the Liſbon earthquakes.

19. The earthquake of † London, on the 8th of March 1750, was ſuppoſed to move from eaſt to weſt. I have been credibly informed, that the ſame thing happened in a ſlight ſhock, which was felt there in the laſt century, as the perſon, who told me this, had an opportunity of obſerving; for being, by accident, in a ſcalemaker's ſhop at the time when it happened, he found that all the ſcales vibrated from eaſt to weſt.

20. All the ſhocks that have been lately felt at Brigue in Valais, have likewiſe come from the ſame point of the compaſs, *viz.* the ſouth ‡.

21. *Fifthly*, The great Liſbon earthquake has been ſucceeded by ſeveral local ones ſince, the extent of which has been much leſs.

22. Such were the earthquakes in Switzerland; thoſe on the borders of France and Germany; thoſe in Barbary, &c. ||

---

* See the accounts of them in Philoſ. Tranſ. N° 209. or vol. ii. Lowthorp's Abr. p. 410, &c.

† See Hiſt. and Philoſ. of Earthq. p. 250. or Philoſ. Tranſ. vol. x. Martyn's Abr. Meteorology, paſſim.

‡ See Philoſ. Tranſ. vol. xlix. p. 620. The ſame has been obſerved at Smyrna alſo, ſee Philoſ. Tranſ. N° 495, or Martyn's Abr. vol. x. p. 526.

|| See the accounts of theſe collected together in Philoſ. Tranſ. vol. xlix. or in the Hiſt. and Philoſ. of Earthq.

SECT. II.

23. How well ſoever theſe facts may agree with the ſuppoſition before laid down, That ſubterraneous fires are the cauſe of earthquakes, one doubt, however, may perhaps remain; *viz.* how it is poſſible that fires ſhould ſubſiſt, which have no communication with the outward air? In anſwer to this, I might alledge the example of green plants, which take fire by fermentation, when laid together in heaps; where the admiſſion of the outward air is ſo far from being neceſſary, that it will effectually prevent their doing ſo. But, to paſs by this, we have many inſtances more immediately to the purpoſe.

24. It can hardly be ſuppoſed, that the fires of the generality of volcanos receive any ſupply of freſh air (for this muſt effectually be prevented by that vapour, which is continually ruſhing out at all their vents), and yet they ſubſiſt, and frequently even increaſe, for many ages. Now, theſe are fires of the very ſame kind with thoſe, which I ſuppoſe to be the cauſe of earthquakes. Other facts, ſtill more expreſly to the purpoſe, are as follow:

25. In the earthquake of the 1ſt of November 1755, we are told, that both ſmoke and light flames were ſeen on the coaſt of Portugal, near Colares; and that, upon occaſion of ſome of the ſucceeding ſhocks, a ſlight ſmell of ſulphur was perceived to accompany a "fog, which came from the ſea, from the ſame "quarter, whence the ſmoke appeared *."

* See Philoſ. Tranſ. vol. xlix. p. 414, &c.

26. In

26. In an account of an earthquake in New England, it is ſaid, that at Newbury, forty miles from Boſton, the earth opened, and threw up ſeveral cart-loads of ſand and aſhes; and that the ſand was alſo ſlightly impregnated with ſulphur, emitting a blue flame, when laid on burning coals *.

27. One of the relaters of the earthquake in Jamaica in 1692, has theſe words: " In Port-Royal, " and in many places all over the iſland, much ſul-" phureous combuſtible matter hath been found " (ſuppoſed to have been thrown out upon the " opening of the earth), which, upon the firſt touch " of fire, would flame and burn like a candle.

28. " St. Chriſtopher's was heretofore much " troubled with earthquakes, which, upon the erup-" tion there of a great mountain of combuſtible mat-" ter, which ſtill continues, wholly ceaſed, and have " never been felt there ſince †."

29. Again, we are told, that, on the 20th November 1720, a burning ‡ iſland was raiſed out of the ſea, near Tercera, one of the Azores, at which place, ſeveral houſes were ſhaken down by an earthquake, which attended the eruption of it. This iſland was about three leagues in diameter, and nearly round; from whence it is manifeſt, that the quantity of pumice ſtones and melted matter, which muſt have been requiſite to form it, was amazingly great:

* See Philoſ. Tranſ. N° 409. or vol. vi. part ii. p. 201. Eames's Abr.

† See Philoſ. Tranſ. N° 209. or vol. ii. p. 418. Lowthorp's Abr.

‡ See Philoſ. Tranſ. N° 372. or vol. vi. part ii. p. 203. Eames's Abr.

in

in all probability, it muſt have far exceeded all that has been thrown out of Ætna and Veſuvius together within the laſt two thouſand years. This may ſerve to ſatisfy us, that the fire which occaſioned all this, muſt have ſubſiſted for many years, not to ſay ages, and this without any communication with the external air. It is worth obſerving, that * ſeveral inſtances of this kind have happened amongſt the Azores. There are beſides many marks of ſubterraneous fires about theſe iſlands, ſeveral places ſending up ſmoke or flames. Theſe iſlands are alſo ſubject to violent and frequent earthquakes.

30. We have more inſtances to the ſame purpoſe, near the iſland of Santerini in the Archipelago, where there have been ſeveral little iſlands raiſed out of the ſea by a ſubmarine volcano. The eruption of one of theſe in the year 1708, with all the circumſtances that attended it, we have a very good account of in the † Philoſophical Tranſactions. It was raiſed in a place where the ſea had been formerly 100 fathoms deep, and was attended with earthquakes before it ſhewed itſelf above water, as well as after. It is reported, that the iſland of Santerini itſelf was originally raiſed out of the ſea in the ſame manner; but, be that as it will, we have certain accounts of new iſlands raiſed there, or additions made to the old ones, from time to time, for above 1900 years backwards, and there have always been earthquakes at the time of theſe eruptions.

* See Hiſt. and Philoſ. of Earthquakes, under the titles Azores, Iſlands raiſed, *&c.*

† See N° 314, 317, and 332. or vol. v. part ii. p. 196. Jones's Abr.

31. An-

31. Another example of the ſame kind happened at * Manila, one of the Philippine iſlands, in the year 1750. This alſo was attended with violent earthquakes, to which that iſland, as well as the reſt of the Philippines, is very much ſubject.

32. We may add to theſe, the many inſtances of vaſt quantities of † pumice ſtones, which have been ſometimes found floating upon the ſea, at ſo great a diſtance from the ſhore, as well as from any known volcano, that there can be little doubt of their being thrown up by fires ſubſiſting under the bottom of the ocean.

33. From theſe inſtances, we may, with great probability, conclude, that the fires of volcanos produce earthquakes: I do not, however, ſuppoſe, that the earthquakes, which are frequently felt in the neighbourhood of volcanos, are owing to the fires of thoſe volcanos themſelves; for volcanos, giving paſſage to the vapours that are there formed, ſhould rather prevent them, as in the inſtance at St. Chriſtopher's, before-mentioned.

34. We alſo meet with frequent inſtances confirming the ſame thing amongſt the Andes. Antonio d'Ulloa (ſpeaking of what happens amongſt theſe mountains) ſays, "Experience ſhews us, that, upon "the freſh breaking out of any volcano, it occaſions "ſo violent a ſhock to the earth, that all the villages, "which are near it, are overthrown and deſtroyed,

* See Philoſ. Tranſ. vol. xlix. p. 459.

† See Philoſ. Tranſ. N° 372. or vol. vi. part ii. p. 204. and N° 402. or vol. vii. part ii. p. 43. Eames's Abr.

" as it happened in the case of the mountain * Carguayraso. This shock, which we may, without the least impropriety, call an earthquake, is seldom found to accompany the eruptions, after an opening is once made; or, if some small trembling is perceived, it is very inconsiderable; so that, after the volcano has once found a vent, the shocks cease, notwithstanding the matter of it continues to be on fire." The greater earthquakes, therefore, seem rather to be occasioned by other fires, that lie deeper in the same tract of country; and the eruptions of volcanos, which happen at the same time with earthquakes, may, with more probability, be ascribed to those earthquakes, than the earthquakes to the eruptions, whenever, at least, the earthquakes are of any considerable extent. If this don't appear sufficiently manifest at present, it will, perhaps, be better understood, by applying to the present purpose, what will be said hereafter concerning local earthquakes.

* It does not appear altogether certain, from the expression made use of in the French translation (from whence I have taken this), that Carguayraso might not have been a volcano in former times, which is asserted to have been the case by Monf. Condamine. It is possible also, that the same may be true of those four mentioned in the next article; and, indeed, it is difficult to know it to be otherwise, in any instance, among the Andes, where the volcanos are generally found at inaccessible heights. But allowing, that all these were only old volcanos, which broke out afresh, yet they will serve at least to swell the number of them in the same neighbourhood, as well as to shew us, that there may, very probably, be many more, which lie hid: for these shewed no marks of their existence, till, by their eruption, they melted a vast quantity of snow, with which they were before covered, and which, being reduced to water, did great damage, by overflowing the country round about.

Sect. III.

35. It may be aſked, perhaps, why we ſhould ſuppoſe, that ſeveral ſubterraneous fires exiſt in the neighbourhood of volcanos? In evidence of this, we have frequent inſtances of new volcanos breaking out in the neighbourhood of old ones: Carguayraſo, juſt mentioned, may ſupply us with one example to this purpoſe; and, in the night of the 28th of October 1746, in which Lima and Callao were deſtroyed, no leſs than four * new ones burſt forth in the adjacent mountains.

36. To the ſame purpoſe, we may allege the inſtances of many volcanos lying together in the ſame tract of country: as for example, the many places, "not ſo few as forty," amongſt the Azores, which either do now or have formerly ſent forth ſmoke and flames; the many volcanos alſo amongſt the Andes, already mentioned: thus Ætna, Strombolo, and Veſuvius, I may add Solfatara too, are all in the ſame neighbourhood: and Monſ. Condamine ſays, he has traced † lavas, exactly like thoſe of Veſuvius, all the way from Florence to Naples. In ‡ Iceland alſo, we have, beſides Hæcla, not only ſeveral other volcanos, but alſo a great number of places, that ſend up ſul-

---

* See d'Ulloa's Voyage to Peru, part ii. book i. chap. 7.

† See Phil. Tranſ. vol. xlix. p. 624. All theſe lavas, as well as the volcanos juſt mentioned, lie in a continued line. The ſame thing holds good in the volcanos of the Andes alſo. This is a fact I muſt deſire the reader to attend to, as it ſerves to confirm a very material doctrine, which I ſhall have occaſion to mention hereafter. See art. 44, 45, and 46.

‡ See Horrebow's Natural Hiſtory of Iceland.

phureous

phureous vapours. But the examples of this kind are ſo frequent, that there are few inſtances to be produced of ſingle volcanos, without evident marks, either that there have been others formerly in their neighbourhood, or that there are, at preſent, ſubterraneous fires near them.

37. This frequency of ſubterraneous fires, in the neighbourhood of volcanos, will appear ſtill more probable, if we conſider the internal ſtructure of the earth; and, as it will be neceſſary alſo, in order to underſtand what follows, to know a little more of this matter, than what falls under common obſervation, I ſhall endeavour to give the reader ſome account of it.

38. The earth then (as far as one can judge from the appearances), is not compoſed of heaps of matter caſually thrown together, but of regular and uniform ſtrata. Theſe ſtrata, though they frequently do not exceed a few feet, or perhaps a few inches, in thickneſs, yet often extend in length and breadth for many miles, and this without varying their thickneſs conſiderably. The ſame ſtratum alſo preſerves a uniform character throughout, though the ſtrata immediately next to each other are very often totally different. Thus, for inſtance, we ſhall have, perhaps, a ſtratum of potters clay; above that, a ſtratum of coal; then another ſtratum of ſome other kind of clay; next, a ſharp grit ſand ſtone; then clay again; next, perhaps, ſand ſtone again; and coal again above that; and it frequently happens, that none of theſe exceed a few yards in thickneſs. There are, however, many inſtances, in which the ſame kind of matter is extended to the depth of ſome

hundreds of yards; but in all theſe, a very few only excepted, the whole of each is not one continued maſs, but is again ſubdivided into a great number of thin laminæ, that ſeldom are more than one, two, or three feet thick, and frequently not ſo much.

39. Beſide the horizontal diviſion of the earth into ſtrata, theſe ſtrata are again divided and ſhattered by many perpendicular fiſſures, which are in ſome places few and narrow, but oftentimes many, and of conſiderable width. There are alſo many inſtances, where a particular ſtratum ſhall have almoſt no fiſſures at all, though the ſtrata both above and below it are conſiderably broken: this happens frequently in clay, probably on account of the ſoftneſs of it, which may have made it yield to the preſſure of the ſuperincumbent matter, and fill up thoſe fiſſures which it originally had; for we ſometimes meet with inſtances in mines, where the correſpondent fiſſures in an upper and lower ſtratum are interrupted in an intermediate ſtratum compoſed of clay, or ſome ſuch ſoft matter.

40. Though theſe fiſſures do ſometimes correſpond to one another in the upper and lower ſtrata, yet this is not generally the caſe, at leaſt not to any great diſtance: thoſe clefts, however, in which the larger veins of the ores of metals are found, are an exception to this obſervation; for they ſometimes paſs through many ſtrata, and thoſe of different kinds, to unknown depths.

41. From this conſtitution of the earth, *viz.* the want of correſpondence in the fiſſures of the upper and lower ſtrata, as well as on account of thoſe ſtrata which are little or not at all ſhattered, it will come to paſs,

paſs, that the earth cannot eaſily be ſeparated in a direction * perpendicular to the horizon, if we take any conſiderable portion of it together; but in the horizontal direction, as there is little or no adheſion between one ſtratum and another, it may be ſeparated without difficulty.

42. Thoſe fiſſures which are at ſome depth below the ſurface of the earth, are generally found full of water; but all thoſe that are below the level of the ſea, muſt always be ſo, either from the oozing of the ſea, or rather of the land waters between the ſtrata.

43. The ſtrata of the earth are frequently very much bent, being raiſed in ſome places, and depreſſed in others, and this ſometimes with a very quick aſcent or deſcent; but as theſe aſcents and deſcents, in a great meaſure, compenſate one another, if we take a large extent of country together, we may look upon the whole ſet of ſtrata, as lying nearly horizontally. What is very remarkable, however, in their ſituation, is, that from moſt, if not all, large tracts of high and mountainous countries, the ſtrata lie in a ſituation more inclined to the horizon, than the country itſelf, the † mountainous countries being generally,

* What I ſaid before of thoſe deep clefts, in which metals are found, will not affect this concluſion; for they are conſiderably different from either perpendicular or plane ſections of earth; they are frequently interrupted by ſtrata of clay, or other ſoft matter; and they are, in moſt parts, either filled up with rubbiſh, or with ores and ſpars, that adhere as firmly to the rocks on both ſides, as if they compoſed one continued ſtratum with them.

† It ſeems very probable, from many appearances, not only that the mountainous countries are formed out of the lower ſtrata of the earth, but that ſometimes the higheſt hills in them are formed out

generally, if not always, formed out of the lower ſtrata of earth. This ſituation of the ſtrata may be not unaptly repreſented in the following manner. Let a number of leaves of paper, of ſeveral different ſorts or colours, be paſted upon one another; then bending them up together into a ridge in the middle, conceive them to be reduced again to a level ſurface, by a plane ſo paſſing through them, as to cut off all the part that had been raiſed; let the middle now be again raiſed a little, and this will be a good general repreſentation of moſt, if not of all, large tracts of mountainous countries, together with the parts adjacent, throughout the whole world *.

44. From this formation of the earth, it will follow, that we ought to meet with the ſame kinds of earths, ſtones, and minerals, appearing at the ſurface, in long narrow ſlips, and lying parallel to the greateſt riſe of any long ridges of mountains; and ſo, in fact, we find them. The Andes in South America, as it has been ſaid before, have a chain of volcanos, that extend in length above 5000 miles: theſe volcanos, in all probability, are all derived from the † ſame

---

out of ſtrata ſtill lower than the reſt, which, perhaps, may always be the caſe, where they have volcanos in them. [See a repreſentation of this in the Plate, Fig. 3.] In other inſtances, however, it often happens, that the hills, to which theſe high lands ſerve as a baſe, are not only formed out of the ſtrata next above them, but they ſtand, as it were, in a diſh, as if they had depreſſed the ground, on which they reſt, by their weight.

* Fig. 1. repreſents a ſection of a ſett of ſtrata, lying in the ſituation juſt deſcribed: the ſection is ſuppoſed to be made at right angles to the length of the ridge, and perpendicular to the horizon.

† See the notes to art. 36 and 53. See alſo Fig. 3.

ſtratum. Parallel to the Andes, is the Sierra, another long ridge of mountains, that run between the Andes and the ſea; and " theſe two ridges of mountains run " within ſight of one another, and almoſt equally, " for above a thouſand leagues together *," being each, at a medium, about twenty leagues wide. The gold and ſilver mines wrought by the Spaniards, are found in a tract of country parallel to the direction of theſe, and extending through a great part of the length of them.

45. The ſame thing is found to obtain in North America alſo. The great lakes, which give riſe to the river St. Laurence, are kept up by a long ridge of mountains, that run nearly parallel to the eaſtern coaſt. In deſcending from theſe towards the ſea, the ſame ſets † of ſtrata, and in the ſame order, are generally met with throughout the greateſt part of their length.

46. In Great Britain, we have another inſtance to the ſame purpoſe, where the direction of the ridge varies about a point from due north and ſouth, lying nearly from ‡ N. by E. to S. by W. There are many more inſtances of this to be met with in the world, if we may judge from circumſtances, which make it highly probable, that it obtains in a great number of places, and in ſeveral they ſeem to put it almoſt out of doubt.

47. The reader is not to ſuppoſe, however, that, in any inſtances, the higheſt riſe of the ridge, and

* See Acoſta's Natural Hiſtory of the Indies.

† See Lewis Evans's Map and Account of North America.

‡ Of this I could give many undoubted proofs, if it would not too far exceed the limits of my preſent deſign, and which, for that reaſon, I am obliged to omit.

the inclination of the ſtrata from thence to the countries on each ſide, is perfectly uniform; for they have frequently very conſiderable inequalities, and theſe inequalities are ſometimes ſo great, that the ſtrata are bent for ſome ſmall diſtance, even the contrary way from the general inclination of them. This often makes it difficult to trace the appearance I have been relating, which, without a general knowlege of the foſſil bodies of a large tract of country, it is hardly poſſible to do.

48. At conſiderable diſtances from large ridges of mountains, the ſtrata, for the moſt part, aſſume a ſituation nearly level; and as the mountainous countries are generally formed out of the lower ſtrata, ſo the more level countries are generally formed out of the upper ſtrata of the earth.

49. Hence it comes to paſs, that, in countries of this kind, the ſame ſtrata are found to extend themſelves a great way, as well in breadth as in length: we have an inſtance of this in the chalky and flinty countries of England and France, which (excepting the interruption of the Channel, and the clays, ſands, *&c.* of a few counties) compoſe a tract of about three hundred miles each way.

50. Beſides the raiſing of the ſtrata in a ridge, there is another very remarkable appearance in the ſtructure of the earth, though a very common one; and this is what is uſually called by miners, the trapping down of the ſtrata; that is, the whole ſet of ſtrata on one ſide a cleft are ſunk down below the level of the correſponding ſtrata on the other ſide. If, in ſome caſes, this difference in the level of the ſtrata, on the different ſides of the cleft, ſhould be

very

very conſiderable, it may have a great effect in producing ſome of the ſingularities of particular earthquakes *.

PART II.

51. IN the former part of this eſſay, I have recounted ſome of the principal appearances of earthquakes, as well as thoſe particulars in the ſtructure of the earth, upon which I ſuppoſe theſe appearances to depend. From what has been already ſaid, I think it is ſufficiently manifeſt, that, in ſome inſtances at leaſt, earthquakes are actually produced by ſubterraneous fires; it now, therefore, remains to be ſhewn, how all the appearances above-recited, as well as many other minuter circumſtances attending earthquakes, may be accounted for from the ſame cauſe.

SECT. I.

52. The returns of earthquakes in the ſame places, either at ſmall or large intervals of time, are very conſiſtent with the cauſe aſſigned: ſubterraneous fires, from their analogy to volcanos, might reaſonably be ſuppoſed to ſubſiſt for many ages, though we had not thoſe inſtances † already mentioned, which put the

* Fig. 2. repreſents a ſection of the ſtrata trapping down after the manner juſt deſcribed. The ſection is ſuppoſed to be made perpendicularly to the horizon, and at right angles to the direction of the cleft: an inſtance of this kind, amongſt the coal miners of Mendip in Somerſetſhire, is mentioned in the Philoſ. Tranſ. See the account of it, together with a drawing, in N° 360. or Jones's Abr. vol. iv. part ii. p. 260.

† See art. 28 to 32 incluſive.

matter out of doubt. And, as it frequently happens, that volcanos rage for a time, and then are quiet again for a number of years; ſo we ſee earthquakes alſo frequently repeated for ſome ſmall time, and then ceaſing again for a long term, excepting, perhaps, now and then ſome ſlight ſhock. And this analogy between earthquakes, and the effects of volcanos, is ſo great, that I think it cannot but appear ſtriking to any one, who will read the accounts of both, and compare them together. The raging of volcanos is not one continued and uniform effect; but an effect, that is repeated at unequal intervals, and with unequal degrees of force: thus, for inſtance, we ſhall have, perhaps, two or three blaſts diſcharged from a volcano, ſucceeding one another at the interval of a few ſeconds only: ſometimes the intervals are of a quarter of an hour, an hour, a day, or perhaps ſeveral days. And as theſe intervals are very unequal, ſo is the violence of the blaſts alſo: ſometimes ſtones, *&c.* are thrown, by theſe blaſts, to the diſtance of ſome miles; at other times, perhaps, not to the diſtance of a hundred yards. The ſame difference is obſerved in the intervals and violence of the ſhocks of earthquakes, which are repeated at ſmall intervals for ſome time.

SECT. II.

53. The great frequency of earthquakes in the neighbourhood of burning mountains, is a ſtrong argument of their proceeding from a cauſe of the ſame kind: and the analogy of ſeveral volcanos lying together in the ſame tract of country, as well as new ones breaking out in the neighbourhood of old ones, tends

tends greatly to confirm this opinion; but what makes it ſtill the more probable, is that peculiarity in the ſtructure of the earth, already mentioned. I obſerved before, that the ſame ſtrata are generally very extenſive, and that they commonly lie more inclining from the mountainous countries, than the countries themſelves: theſe circumſtances make it very probable, that thoſe * ſtrata of combuſtible materials, which break

* It has been imagined by ſome authors, that volcanos are produced by the pyrites of veins, and that they do not owe their origin to the matter of ſtrata. In order to prove this, it is alleged, that volcanos are generally found on the tops of mountains, and that thoſe are the places in which veins of pyrites are generally lodged. This argument being taken from obſervations that have their foundation in nature, ought not to go unanſwered. In the firſt place, then, the pyrites of veins, or fiſſures, are not found in ſufficient quantities, or extending to a ſufficient breadth, to be ſuppoſed capable of producing the fires of volcanos: it very rarely happens, that we meet with a vein or fiſſure five or ſix yards wide; and when we meet with ſuch an one, yet, perhaps, not a twentieth part of it at moſt ſhall be filled with pyrites; but the fires of volcanos, inſtead of being long and narrow, as if the matter that ſupplied them was depoſited in veins, are generally round, and of far greater breadth than veins can be ſuppoſed to be. Monſ. Bouguer ſays, that the mouth of the volcano Cotopaxi is, at this time, five or ſix hundred fathoms wide; [ſee Hiſt. and Phil. of Earthquakes, p. 195.] and the burning iſland that was raiſed out of the ſea near Tercera, as before-mentioned, was almoſt three leagues in diameter, and nearly round. [See art. 29.]

Beſides this, it is very difficult to conceive how any matters lodged in veins can ever take fire; for, excepting where the veins are extremely narrow, they are almoſt always drowned in a very great quantity of water, which has free acceſs to every part of them: neither are the pyrites of veins, by any means, ſo apt to take fire of themſelves, as thoſe of ſtrata; and if, indeed, there are any of them that will do ſo, yet they are but few in compariſon of thoſe which will not: all thoſe, which, beſide iron and ſulphur, contain copper, or arſenic, even in a very ſmall proportion, are not at

break out in volcanos on the tops of the hills, are to be found at a confiderable depth under ground in the level and low countries near them. If this fhould be the cafe,

---

at all fubject to inflame of themfelves. On the other hand, moft of the pyrites of ftrata, if not all of them, have this property more or lefs. There are alfo two forts of ftrata, in which pyrites are lodged in the greateft abundance, that have the fame property, and that frequently in as great a degree as themfelves: thefe are coals and aluminous earths, or fhale. There are fome kinds of both thefe, that, upon being expofed to the external air for a few months, will take fire of themfelves, and burn. Thefe two forts of ftrata are alfo near akin to each other; they are generally found to accompany each other; they are both of them generally intermixed with, or accompanied by ftrata of iron ore; and they both of them, for the moft part, either contain, or are lodged amongft, the remains of vegetable bodies; and thefe remains of vegetable bodies, in the aluminous earths, are frequently either wholly, or in part, converted into pyrites, or coal, or both. Numberlefs, inftances of this are to be met with in the aluminous fhale of Whitby and other places.

It is very probable, that to fome ftratum of this kind the fires of volcanos are owing; and this feems to be confirmed by the fimilarity of the materials, which are thrown up or fublimated by the fires of volcanos, to the matter of the aluminous earths. Solfatara produces fulphur, alum, and fal ammoniac. The two former of thefe are very éafily to be obtained from the aluminous earths, and, I fuppofe, the latter alfo; at leaft it is procurable from the foot of common foffil coals, and probably, therefore, from the foot of that coaly matter which is intermixed with fuch earths.

The aluminous earths, moreover, not only have feveral ftrata of iron ore lying in them, but they alfo contain a confiderable proportion of iron in their compofition. In correfpondence to this, we find the lavas of volcanos, and other matters thrown out from thence, frequently containing a great deal of iron, the fmall duft of them readily adhering to the magnet.

As to the pyrites of veins, I much doubt whether they ever contain alum, or fal ammoniac; at leaft they are very rarely found to contain either the one or the other.

and

and if the ſame * ſtrata ſhould be on fire in any places under ſuch countries, as well as on the tops of the hills, all vapours, of whatſoever kind, raiſed from theſe fires, muſt be pent up, unleſs ſo far as they can open themſelves a paſſage between the ſtrata; whereas the vapours raiſed from volcanos find a vent, and are diſcharged in blaſts from the mouths of them. Now, if, when they find ſuch a vent, they are yet capable of ſhaking the country to the diſtance of ten or twenty miles round, what may we not expect from them, when they are confined? We may form ſome idea of the force and quantity of theſe vapours from their effects: it is no uncommon thing to ſee them throw up, at once, ſuch clouds of ſand, aſhes, and pumice ſtones, as are capable of darkening the whole air, and covering the neighbouring country with a ſhower of duſt, *&c.* to ſome miles diſtance. great ſtones alſo, of ſome tons weight, are often thrown to the diſtance of two or three miles by theſe exploſions: and Monſ. Bouguer tells us, that he met with ſtones

---

* It may be aſked, perhaps, why a ſtratum liable to take fire in ſome places, ſhould not take fire throughout the whole extent of it? In anſwer to this, it may be ſaid, that the ſame ſtratum may differ a little in the richneſs of its combuſtible principles in different places; or, perhaps, the frequency of the fiſſures, either in the combuſtible ſtratum itſelf, or the ſtratum next to it, may let in ſo much water, as to prevent its taking fire, excepting in a few places; but, if this once happens, the fire will not eaſily be put out again, but it will ſpread itſelf, notwithſtanding the fiſſures that lie in its way, though they are filled with water; for the matter on fire will be, in ſome degree at leaſt, in a fluid ſtate; and, for this reaſon, it muſt neceſſarily expel the water from the fiſſures, both on account of the extenſion of its own dimenſions by the heat, and of the weight of the ſuperincumbent earth, which, preſſing it, will make it ſpread laterally.

in South America, of eight or nine feet diameter, that had been thrown from the volcano Cotopaxi, by one of these blasts, to the distance of more than * three leagues.

54. If we suppose that these vapours, when pent up, are the cause of earthquakes, we must naturally expect, from what has been just said, that the most extensive earthquakes should take their rise from the level and low countries; but more especially from the sea, which is nothing else than waters covering such countries. Accordingly we find, that the great earthquake of the 1st November 1755, which was felt at places near three thousand miles distant from each other, took its rise from under the sea; this is manifest, from that wave which accompanied it, as shall be shewn hereafter. The same thing is to be understood of the earthquake that destroyed Lima in the year 1746, which, it has been said, was felt as far as Jamaica; and, as it was more violent than the Lisbon earthquake, so, if this be true, it must, in all probability, have been more extensive also. There have been many other very extensive earthquakes in South America: Acosta says, that they have been often known to extend themselves one, two, or three hundred, and some even five hundred leagues, along the coast. These have been generally, if not always, attended with waves from the sea; but any minuter

* See Hist. and Philos. of Earthq. p. 195. Don Antonio d'Ulloa, an author of great veracity, speaking of the same thing, says, that "the whole plain [near Latacunga] is full of large pieces of rocks, "some of them thrown from the volcano Cotopaxi, by one of its "eruptions, to the distance of five leagues." See his Voyage to Peru, part i. book vi. chap. 1.

circum-

circumſtances accompanying them are not related. Indeed it is hardly to be expected that they ſhould be obſerved, much leſs that they ſhould be related, when they happened in a country ſo thinly inhabited, and where one may reaſonably ſuppoſe, that, in general, only the groſſer and more violent effects would be taken notice of.

SECT. III.

55. I have ſaid before, that I imagined earthquakes were cauſed by vapours raiſed from waters ſuddenly let out upon ſubterraneous fires. It is not eaſy to find any other cauſe capable of producing ſuch ſudden and violent effects, or of raiſing ſuch an amazing quantity of vapour in ſo ſmall a time. That the blaſts, diſcharged from volcanos, are always produced from this cauſe, is highly probable; that they are often ſo, cannot admit of the leaſt doubt. There can be no doubt, that conſiderable quantities of water muſt be often let out upon the fires of theſe volcanos, and whenever this happens, it will be immediately raiſed by the heat of them into a vapour, whoſe elaſtic force is capable of producing the moſt violent effects *.

56. Both

* There are many effects produced by the vapour of water, when intenſely heated, which make it probable, that the force of gunpowder is not near equal to it. The effects of an exceeding ſmall quantity of water, upon which melted metals are accidentally poured, are ſuch, as, I think, could in no wiſe be expected from the like quantity of gunpowder. Founders, if they are not careful, often experience theſe effects to their coſt. An accident of this kind happened about forty years ſince, at the caſting of two braſs cannon at Windmill-hill, Moorfields. " The " heat

56. Both the tremulous and wave-like motion obſerved in earthquakes, may be very well accounted for

" heat of the metal of the firſt gun drove ſo much damp into the " mould of the ſecond, which was near it, that as ſoon as the " metal was let into it, it blew up with the greateſt violence, tear- " ing up the ground ſome feet deep, breaking down the furnace, " untiling the houſe, killing many ſpectators on the ſpot, with the " ſtreams of melted metal, and ſcalding many others in a moſt mi- " ſerable manner." [See the note at the end of proceſs 44th of the Engliſh tranſlation of Cramer's Art of aſſaying Metals.]

Other inſtances of the violence of vapours raiſed from water, are frequently to be met with: one of Papin's digeſters being placed between the bars of a grate, where there was a fire, was, after ſome time, burſt by the violence of the ſteam, the fire was all blown out of the grate, and a piece of the digeſter was driven againſt the leaf of a ſtrong oak table, which it broke to pieces. [See Philoſ. Tranſ. N° 454. or Martyn's Abr. vol. viii. p 465.] The marquis of Worceſter alſo, in his Century of Inventions, tells us, that he burſt a cannon by the ſame means.

It has been ſometimes imagined, that the vapours, which occaſion earthquakes, were of the ſame kind with thoſe fulminating damps, of which we often meet with inſtances in coal mines. Now, there are ſeveral things which make it very probable, that this is not the caſe: it is true, the force of ſuch vapours is very great; we have had inſtances, where large beams of timber have been thrown to the diſtance of an hundred yards by them: [ſee Philoſ. Tranſ. N° 136. or vol. ii. p. 381. Lowthorp's Abr.] but what is this to the force of that vapour, which could throw ſtones of twenty or thirty ton weight to the diſtance of three leagues? Nor, indeed, is it at all probable, that any vapour, already in the form of a vapour, can, by ſuddenly taking fire, increaſe its dimenſions ſo much, as to produce that immenſe quantity of motion, which we obſerve in ſome earthquakes. but this is rather to be expected from ſome ſolid body, ſuch as water, which is capable of being converted, and that almoſt inſtantly, into one of the lighteſt, and perhaps one of the moſt elaſtic, vapours in the world. Air, when heated to the greateſt degree that it is capable of receiving from the hotteſt fires we can make, acquires a degree of elaſticity about five times as great as that of common air: the vapour of gun-for from ſuch a vapour. In order to trace a little more particularly the manner in which theſe two motions

---

gunpowder, whilſt it is inflamed, has alſo about five times the elaſtic force which it has when cold. [See Robins's excellent tract on Gunnery.] Now, if we ſuppoſe a fulminating damp, of any kind, to increaſe its elaſticity, when inflamed in the ſame proportion, this will be abundantly ſufficient to make it produce any effects, which we have ever ſeen produced by any of the damps of mines, &c. And, indeed, whoever carefully examines the effects, either of the damps of mines, or of thoſe fulminating damps, that are raiſed from ſome metals, when in fuſion, or when they are diſſolving in acids, will rather be inclined to think, that the force of inflamed vapours is ſo far from exceeding the proportion of five to one, that it falls conſiderably ſhort of it.

But though we ſhould ſuppoſe that this proportion holds good, where ſhall we find a place capable of containing a ſufficient quantity of ſuch a vapour, to produce the great effects of earthquakes? It will be ſaid, perhaps, in ſubterraneous caverns. To this we may anſwer, that he, who is but moderately acquainted with the ſtructure of the earth, and the materials of which it is compoſed, will be little inclined to allow of any great or extenſive caverns in it. But, though this ſhould be admitted, how can it come to paſs that theſe caverns ſhould not be filled with water? If it is alleged, that the water is expelled, as the vapour is formed, why ſhould not the vapour, as it is ſuppoſed to be the lighter, be expelled rather than the water, by the ſame paſſages by which the water is to be expelled? But let us ſuppoſe this difficulty alſo to be got over, and the water to be removed, and we ſhall then have a gage for the denſity of the vapour; for it muſt be juſt ſufficient to make it capable of ſuſtaining a column of water, whoſe height is equal to that of the ſurface of the ſea above the bottom of the cavern, in which the vapour is ſuppoſed to be contained. Now, ſince the mean weight of earth, ſtones, &c. is not leſs than two and a half times the weight of water, this vapour muſt be increaſed to two and half times its original elaſticity, before it can, in any wiſe, raiſe the earth above it; and if we ſuppoſe it to be increaſed to five times its original elaſticity, it will then be no more than twice able to do ſo; in which caſe, ſo much vapour only can be diſcharged from the cavern, to produce an earthquake, as is equal to the content

motions will be brought about, let us ſuppoſe the roof over ſome ſubterraneous fire to fall in. If this ſhould be the caſe, the earth, ſtones, &c. of which it was compoſed, would immediately ſink in the melted matter of the fire below: hence all the water contained in the fiſſures and cavities of the part falling in, would come in contact with the fire, and be almoſt inſtantly raiſed into vapour. From the firſt effort of this vapour, a cavity would be formed (between the melted matter and ſuperincumbent earth) filled with vapour only, before any motion would be perceived at the ſurface of the earth: this muſt neceſſarily happen, on account of the * compreſſibility of

---

content of the cavern: and what muſt the ſize of that cavern be, which could contain vapour enough to produce the earthquake of the 1ſt of November 1755, in which an extent of earth of near three thouſand miles diameter was conſiderably moved? or how can we ſuppoſe, that the roof of ſuch a cavern, when ſo violently ſhaken, ſhould avoid falling in? eſpecially, as it is hardly to be ſuppoſed, that any inflamed vapour whatſoever ſhould be able to move the earth over theſe caverns, if they lay at any great depth, ſince the weight of leſs than three miles depth of earth is capable of retaining the inflamed vapour of gunpowder within the original dimenſions of the gunpowder itſelf; and common air, compreſſed by the ſame weight (ſuppoſing the known law of its compreſſion to hold ſo far), would be of greater denſity than water.

We may aſk ſtill farther, whence ſuch vaſt quantities of vapour ſhould be formed, or what ſources they muſt be, which would not be exhauſted (if they were not again repleniſhed) by a very few repetitions of ſuch immenſe diſcharges.

* The compreſſibility and elaſticity of the earth, are qualities which don't ſhow themſelves in any great degree in common inſtances, and therefore are not commonly attended to. On this account it is, that few people are aware of the great extent of them, or the effects that may ariſe from them, where exceeding large quantities of matter are concerned, and where the compreſ- ſive

of all kinds of earth, ſtones, &c. but as the compreſſion of the materials immediately over the cavity, would

---

ſive force is immenſely great. The compreſſibility and elaſticity of the earth may be collected, in ſome meaſure, from the vibration of the walls of houſes, occaſioned by the paſſing of carriages in the ſtreets next to them. Another inſtance to the ſame purpoſe, may be taken from the vibrations of ſteeples, occaſioned by the ringing of bells, or by guſts of wind: not only ſpires are moved very conſiderably by this means, but even ſtrong towers will, ſometimes, be made to vibrate ſeveral inches, without any disjointing of the mortar, or rubbing of the ſtones againſt one another. Now, it is manifeſt, that this could not happen, without a conſiderable degree of compreſſibility and elaſticity in the materials, of which they are compoſed: and if ſuch ſmall things as the weight of ſteeples, and the motion of bells in them, or a guſt of wind, are capable of producing ſuch effects, what may we not expect from the weight of great depths of earth? There are ſome circumſtances, which ſeem to make it not altogether improbable, that the form and internal ſtructure of the earth depend, in a great meaſure, upon the compreſſibility and elaſticity of it. There are ſeveral things that ſeem to argue a conſiderably greater denſity in the internal, than the external part of the earth; and why may not this greater denſity be owing to the compreſſion of the internal parts ariſing from the weight of the ſuperincumbent matter, ſince it is probable, that the matter, of which the earth is compoſed, is pretty much of the ſame kind throughout? There is a ſtill ſtronger argument for the earth's owing its form, in ſome meaſure, to the ſame cauſe; for it is found to be higher [ſee the French accounts of the meaſures of a degree of the meridian in France, Sweden, and America] at the æquator, than at the poles, in a greater proportion than it would be on account of the centrifugal force, if it was of uniform denſity; but, if we ſuppoſe the earth to be of leſs denſity in an æquatorial diameter than in the axis, the whole will then be eaſily accounted for, from the riſing of the earth a little by its elaſticity, the weight being in part taken off by the diurnal rotation: and that the earth is really a little denſer in the axis, than in the æquatorial diameter, ſeems highly probable, from the experiments of pendulums compared with aſtronomical obſervations; for the forms of the earth derived from theſe, cannot be

reconciled

would be more than ſufficient to make them bear the weight of the ſuperincumbent matter, this compreſſion muſt be propagated on account of the elaſticity of the earth, in the ſame manner as a pulſe is propagated through the air; and again the materials immediately over the cavity, reſtoring themſelves beyond their natural bounds, a dilatation will ſucceed to the compreſſion; and theſe two following each other alternately, for ſome time, a vibratory motion will be produced at the ſurface of the earth. If theſe alternate dilatations and compreſſions ſhould ſucceed one another at very ſmall intervals, they would excite a like motion in the air, and thereby occaſion a conſiderable noiſe. The noiſe that is uſually obſerved to precede or accompany earthquakes, is probably owing partly to this cauſe, and partly to the grating of the parts of the earth together, occaſioned by that wave-like motion before-mentioned.

57. After the water, that firſt came in contact with the fire, has formed a cavity, all the reſt of the water contained in the fiſſures, immediately communicating with the hollow left by the part that fell in,

reconciled with each other, but upon this ſuppoſition. [See Maclaurin's Fluxions, art. 681, &c.] It appears, from ſome late and accurate obſervations, that the æquatorial parts of the planet Jupiter alſo, as well as thoſe of the earth, are a little higher than they would be, if their riſe was owing to the centrifugal force, and he was of uniform denſity; but if we ſuppoſe him to be of leſs denſity in the æquatorial, than the polar regions, then the form may be ſuch as he would aſſume from the reſpective gravitation of the ſeveral parts; and any fluid like our ocean, would not overflow the polar parts, (which, upon any other ſuppoſition, it muſt neceſſarily do) but would follow his general form, as our ocean does that of the earth.

muſt

muſt run out upon the fire, the ſteam taking its place. From hence may be generated a vaſt quantity of vapour, the effects of which ſhall be conſidered preſently. This ſteam will continue to be generated, ſuppoſing the fire to be ſufficiently great, till the fiſſures before-mentioned are evacuated, or till the water begins to flow very ſlowly; when the ſteam already formed will be removed by the elaſticity of the earth, which will again ſubſide, and, preſſing upon the ſurface of the melted matter, will force it up a little way into all the clefts, by which the water might continue to flow out. By this means, all communication between the fire and the water will be prevented, excepting at theſe clefts, where the water, dripping ſlowly upon the melted matter, will gradually form a cruſt upon it, that will ſoon ſtop all farther communication in theſe places likewiſe; and the fiſſures, that had been before evacuated, will be again gradually repleniſhed by the oozing of the water between the ſtrata.

58. As a ſmall quantity of vapour almoſt inſtantly generated at ſome conſiderable depth below the ſurface of the earth, will produce a vibratory motion, ſo a very large quantity (whether it be generated almoſt inſtantly, or in any ſmall portion of time) will produce a wave-like motion. The manner in which this wave-like motion will be propagated, may, in ſome meaſure, be repreſented by the following experiment. Suppoſe a large cloth, or carpet, (ſpread upon a floor) to be raiſed at one edge, and then ſuddenly brought down again to the floor, the air under it, being by this means propelled, will paſs along, till it eſcapes at the oppoſite ſide, raiſing the cloth

cloth in a wave all the way as it goes. In like manner, a large quantity of vapour may be conceived to raiſe the earth in a wave, as it paſſes along between the ſtrata, which it may eaſily ſeparate in an horizontal direction, there being, as I have ſaid before, little or no coheſion between one ſtratum and another. The part of the earth that is firſt raiſed, being bent from its natural form, will endeavour to reſtore itſelf by its elaſticity, and the parts next to it beginning to have their weight ſupported by the vapour, which will inſinuate itſelf under them, will be raiſed in their turn, till it either finds ſome vent, or is again condenſed by the cold into water, and by that means prevented from proceeding any farther.

59. If a large quantity of vapour ſhould continue to be generated for ſome time, ſeveral waves might be produced by it; and this would be, in ſome meaſure, the caſe, if the quantity at firſt generated was exceedingly great, though the whole of it was generated in leſs time, than whilſt the motion was propagated through the diſtance between two waves.

60. Theſe waves muſt riſe the higher, the nearer they are to the place from whence they have their ſource; but, at great diſtances from thence, they may riſe ſo little, and ſo ſlowly, as not to be perceived, but by the motions of waters, hanging branches in churches, *&c.*

61. The vibratory motion occaſioned by the firſt impulſe of the vapour, will be propagated through the ſolid parts of the earth, and therefore, it will much ſooner become too weak to be perceived, than the wave-like motion; for this latter, being occaſioned by the vapour inſinuating itſelf between the

ſtrata,

ſtrata, may be propagated to very great diſtances; and even after it has ceaſed to be perceived by the the ſenſes, it may ſtill diſcover itſelf by the appearances before-mentioned.

SECT. IV.

62. All earthquakes derived from the ſame ſubterraneous fire, muſt come to the ſame place in the ſame direction; and thoſe only which are derived from different fires, will come from different points of the compaſs; but as, in all probability, it ſeldom happens that earthquakes, cauſed by different fires, affect the ſame plaee, we therefore find in general, that they come from the ſame quarter: it is not, however, to be ſuppoſed, that this ſhould always be the caſe, for it will, probably, ſometimes happen to be otherwiſe: and this is to be expected in ſuch places as are ſituated in the neighbourhood of ſeveral ſubterraneous fires; or where, being ſubject to the ſhocks of ſome local earthquake of ſmall extent, they now and then are affected by an earthquake, produced by ſome more diſtant, but much more conſiderable cauſe. Of this laſt caſe, we ſeem to have had ſome inſtances in the earthquake of the 1ſt of November 1755, and thoſe local ones, before-mentioned, which ſucceeded it.

63. As we may reaſonably infer from many earthquakes coming to the ſame place, from the ſame point of the compaſs, that they are all derived from the ſame cauſe, and that a permanent one; ſo we may reaſonably infer the ſame thing alſo, from their being propagated with the ſame velocity; but this argument will ſtill come with the greater force, if it be

conſidered,

confidered, that the velocity of any vapour, which infinuates itfelf between the ftrata of the earth, depends upon the depth of it below the furface; for the deeper it lies, the greater will be its * velocity. We may therefore conclude, from the famenefs of the velocity of the earthquakes of the fame place, that the caufe of them lies at the fame depth; and from the inequality of the velocity of the earthquakes of different places, that their caufes lie at different depths. Both thefe are perfectly confiftent with the fuppofition, that earthquakes owe their origin to fubterraneous fires, fince the ftrata in which thefe fubfift, may be eafily conceived to lie at different depths in different parts of the world.

Sect. V.

64. From the fame caufe, we may eafily account for thofe local earthquakes, which fucceed the greater and more extenfive ones. If there are many fubterraneous fires fubfifting in different parts of the world, the vapour coming from one fire may very well be fuppofed, as it paffes, to difturb the roof over fome other fire, and, by that means, occafion earthquakes by the falling in of fome part of it: and this may be the cafe, in fome meafure, even where the vapour paffes at fome fmall diftance over the fire; but it will be moft likely to take place, where the vapour either

* The velocity of fuch a vapour, depending intirely upon the elafticity of the earth which is over it, will be, *cæteris paribus*, (if I am not miftaken) in the ratio of the depth below the furface. This feems to follow from a known law of all elaftic bodies, according to which they tend to return to their ftate of reft, when either dilated or compreffed, with forces proportionable to the quantity by which they differ from their natural bounds.

paffes

paſſes at ſome diſtance under it, or between the ſtratum, in which the fire lies, and that next above or below it.

PART III.

SECT. I.

65. IN the former part of this tract, I ſuppoſed a part of the roof over ſome ſubterraneous fire to fall in: this is an event that cannot happen merely accidentally; for ſo long as the roof reſts on the matter on fire, no part of it can fall in, unleſs the matter below could riſe and take its place: now, it is very difficult to conceive how this ſhould happen, unleſs it was to riſe by ſome larger paſſages than the ordinary fiſſures of the earth, which ſeem much too narrow for that purpoſe; for, beſides that the melted matter cannot be ſuppoſed to have any very great degree of fluidity, it muſt neceſſarily have a hard cruſt formed upon it, at all the fiſſures, by the long continued contact of the water contained in them: theſe impediments ſeem too great to be overcome by the difference of the ſpecific gravities of the part that is to fall in, and the melted matter, which is the only cauſe that can tend to make it deſcend; the manner therefore, in which, I ſuppoſe, this event may be brought about, is as follows:

66. The matter of which any ſubterraneous fire is compoſed, muſt be greatly * extended beyond its original

* As all bodies we are acquainted with are liable to be extended by heat, there can be no doubt of its being ſo in this caſe

likewiſe;

ginal dimensions by the heat. As this will be brought about gradually, whilst the matter spreads itself, or grows hotter, the parts over the fire will be gradually raised and bent; and this bending will, for some time, go on without any other consequence; but, as the fire continues to increase, the earth will at last begin to be raised somewhat beyond the limits of it. By this means, an annular space will be formed at the edges next to the fire, and surrounding it, a vertical section of which space, through a diameter of the fire, will be two long triangles, the shortest side or base of each lying next the fire, and the two longer sides being formed by the upper and lower strata, which will be separated for a considerable extent, proportionably to the distance through which they are raised from each other *. This space will be gradually

likewise; but the matter of subterraneous fires is yet much more extended, than those bodies which are only capable of being melted into a solid glass, if we may judge of it from what we see of volcanos; for the lavas, sciari, and pumice stones, thrown out from thence, even after they are cold, are commonly of much less specific gravity, on account of their porous spongy texture, than the generality of earth, stones, *&c.* and they frequently are even lighter than water, which is itself lighter than any known fossil bodies, that compose strata in their natural state.

* In Fig. 4. A is supposed to represent a vertical section of the matter on fire; B B, parts of the same stratum yet unkindled; C C, the two sections of the annular space, (surrounding the fire) which is supposed to be filled with water, as far as the strata are separated; D, the several sets of earth, stones, *&c.* lying over the fire, which are raised a little, and bent, by the expansion of the matter at A. As it is not easy to represent the things above described in their due proportions, it may not be amiss, in order to prevent the figure here given from misleading the reader, to give some random measures of the several parts, such as may probably approach

gradually filled with water, as it is formed, the melted matter being prevented from filling it, by its want of fluidity, as well as on account of the other circumſtances, under which it is to ſpread itſelf; for the lentor and ſluggiſhneſs of this kind of matter is ſuch, that, when ſomewhat cooled on the ſurface by the contact of the air only, it will not flow, perhaps, ten feet in a month, though in a very large body; inſtances of which we have in the lavas of Ætna, Veſuvius, *&c.* It is not to be expected then, that it ſhould ſpread far, when it comes in contact with water at its edges, as ſoon as it is formed, and when it is, perhaps, ſeveral months in acquiring a thickneſs of a few inches; but it muſt, by degrees, form a kind of wall between the fire and the opening into the annular ſpace before deſcribed. This wall will gradually increaſe in height, till it becomes too tall in proportion to its thickneſs, to bear any longer the preſſure of the melted matter; which

---

approach towards thoſe which are ſometimes found in nature: we may ſuppoſe then the ſtratum B to be, perhaps, from ten or twenty to a hundred yards in thickneſs; the greateſt height of the annular ſpace C, next the fire, to be from four or five to ten or fifteen feet, and its greateſt extent, horizontally, from ten or twenty to fifty or ſixty feet; the horizontal extent of the fire at A, may be from half a mile to ten or twenty miles; [See art. 29. and the note to art. 53.] and the thickneſs of the ſuperincumbent matter at D, may be from a quarter or half a mile to two or three miles; the number of the laminæ alſo, into which it is divided, may be many times more than thoſe in the figure. As to the perpendicular fiſſures, they muſt be ſo numerous, and ſo ſmall, in proportion to the other parts, that I choſe rather to leave them, to be ſupplied by the imagination of the reader, than attempt to expreſs them in a manner, that could give no adequate idea of them at all.

muſt neceſſarily happen at laſt, becauſe the thickneſs of it will not exceed a certain * limit.

67. Beſides the giving way of this wall, the fire may undermine the ſpace containing the water, and, by that means, open a communication between them. Let us ſuppoſe one of theſe come to paſs, and the time arrived when the partition begins to yield. If then the water had any way to eſcape readily, the breach would be made, and the melted matter would burſt forth immediately, and flow out in large quantities at once amongſt it; but as this is not the caſe, and it can only eſcape by oozing ſlowly between the ſtrata, and through the fiſſures, the way that it came, the breach will be made gradually, from whence we may account for ſome appearances that have preceded great earthquakes.

68. We are told, that two or three days before an † earthquake in New England, the waters of ſome wells were rendered muddy, and ſtank intolerably:

---

* This limit will depend upon the thickneſs of the matter neceſſary to prevent ſo quick a communication of the heat or cold through it, as that the water ſhould be able to diminiſh the heat of the fire conſiderably. The thickneſs requiſite to do this, is very different in different kinds of bodies. Metals of all kinds tranſmit heat and cold extremely readily; but bricks and vitrified ſubſtances (with which laſt we may claſs the matter under our preſent conſideration) tranſmit them very ſlowly: the walls of the hotteſt of our furnaces, when built of bricks, and eighteen inches thick, will not tranſmit more heat than a living animal can bear without injury, though the fires are continued in them for ever ſo long a time; probably, therefore, if we allow two feet for the thickneſs of the matter, cooled and rendered hard by the contact of the water, we ſhall not underdo it.

† See Philoſ. Tranſ. N° 437. or Martyn's Abridgm. vol. viii. p. 689.

why might not this be occaſioned by the waters contained in the ſpaces before deſcribed, which, being impregnated with ſulphureous ſteams, were driven up, and mixed with the waters of the ſprings? At leaſt, there can be no doubt, by whatſoever means it was brought about, that this phænomenon was owing to the ſame cauſe, already beginning to exert itſelf, which afterwards gave riſe to the ſucceeding earthquake.

69. Something like this happened before the great Liſbon * earthquake of 1755. We are told, that at Colares, about twenty miles from thence, " in the afternoon preceding the 1ſt of November, the water of a fountain was greatly decreaſed: on the morning of the 1ſt of November, it ran very muddy, and after the earthquake, it returned to its uſual ſtate, both in quantity and clearneſs." The ſame author ſays, a little lower, " in the afternoon of the 24th, I was much apprehenſive, that the following days we ſhould have another great earthquake; for I obſerved the ſame prognoſtics as in the afternoon of the 31ſt October; that is," *&c.* " And I farther obſerved, that the water of a fountain began to be diſturbed to ſuch a degree, that in the night it ran of a yellow clay colour; and from midnight to the morning of the 25th, I felt five ſhocks, one of which ſeemed to me as violent as that of the 11th of December."

70. But the moſt extraordinary appearance of any that preceded this earthquake, was that of the agita-

* See Philoſ. Tranſ. vol. xlix, p. 416 and 417.—or Hiſt. and Philoſ. of Earthq. p. 313.

tion

tion of the waters of * Lochneſs, and ſome others of the lochs in Scotland, about half an hour before any motion was felt at Liſbon, notwithſtanding the cauſe of all theſe great effects could not lie far from thence, and, I think, certainly lay to the ſouth of Oporto. Nor is it probable, that there ſhould be any miſtake in the time, not only becauſe the difference is too great, as well as the concurrent teſtimonies too many, to admit of ſuch a ſolution; but becauſe they mention another greater agitation, that happened about an hour and half after the former; which latter agrees with the times, when the agitations of the waters were obſerved in England, if we allow only a proper interval for the motion to be propagated ſo far northward, proportionably to the time it took up in travelling from its original ſource near Liſbon.

71. Theſe appearances ſeem to be connected with that mentioned in the preceding article, and they may both, I think, be accounted for, by ſuppoſing a conſiderable quantity of vapour to be raiſed, whilſt the partition before-mentioned was beginning to give way; during which time, a partial

* See Philoſ. Tranſ. vol. xlix.—or Hiſt. and Philoſ. of Earthq. art. Lochneſs, Lochlommond, &c. The ſame thing alſo ſeems to have taken place in Switzerland; for Monſ. Bertrand ſays, that all the agitations of the waters in the lakes there, which were obſerved on the 1ſt November 1755, happened between nine and ten in the morning; and particularly at lake Leman, he ſays, the agitation happened juſt before ten; which, allowing for the difference of longitude, muſt have been juſt before nine at Liſbon; and, conſequently, if there is no miſtake in the times, all theſe agitations preceded the earthquake, at this laſt place, by near three quarters of an hour. [See *Memoires ſur les tremblemens de Terre*, p. 107 *et* 105.]

communi-

communication between the water and fire would be brought on, and that by degrees only. Hence the vapour, not being produced at once but gradually, might creep * ſilently between the ſtrata, towards that quarter where the ſuperincumbent maſs of earth was lighteſt; and, by this means, ſome places very near the ſource of the vapour might be little, or not at all, affected by it, whilſt others might be greatly affected, though they lay at a great diſtance; and even thoſe places, which lay immediately over the part where the vapour was paſſing, might not perceive any effect, on account of the gentleneſs of the motion, occaſioned by the ſmall quantity of it. This might continue to be the caſe, till it came to ſome country where, the ſet of ſtrata above being much thinner, the vapour would not only be hurried forward, but collected alſo into a much narrower compaſs; and therefore, raiſing the earth more, would produce more ſenſible effects; and this we ought

* Some appearances that have been obſerved in New England ſeem to confirm this, and make it probable, that a ſmall quantity of vapour is often found to creep ſilently between the ſtrata, before a general communication between the water and the fire gives riſe to the greater and more ſenſible effects of earthquakes. See Philoſ. Tranſ. N° 462. or Martyn's Abr. vol. viii. p. 693. where we are told, that, at Newbury, a little before any noiſe or ſhock was perceived, the bricks of an hearth were obſerved to riſe, and, falling down again, to lean another way. In the ſame account, it is alſo ſaid, that " a few minutes before any ſhock came, many people " could foretell it by an alteration in their ſtomachs:" an effect, which ſeems to be of the ſame kind with ſea-ſickneſs, and which always accompanies the wave-like motion of earthquakes, when it is ſo weak, as to be uncertainly diſtinguiſhable.

chiefly

chiefly to expect in the moſt mountainous countries, according to the idea before given of them *.

72. To make this ſomething clearer, let us ſuppoſe, in Fig. 1. the vapour to be paſſing between the ſtrata in the dotted line C, and to go forwards, till it arrives at A: whilſt, then, it paſſes under the deeper parts at E, it will raiſe the earth over it but little, as well becauſe it will be ſpread broader and thinner, as becauſe it will be more compreſſed by the weight of the ſuperincumbent matter; but as it arrives towards A, not only the latter part will be driven forwards with greater velocity, but the foremoſt will travel ſlower, on account of its travelling under a † thinner ſet of ſtrata; and, beſides this, the load being much leſs, it will greatly expand itſelf. From all theſe cauſes taken together, the wave at the ſurface of the earth, occaſioned by the paſſing of the vapour under it, will not only be much higher, but alſo much ſhorter, and, conſequently, the ſides of it, on both theſe accounts, will be much more inclined to the horizon: and, moreover, becauſe the progreſs of the wave will be ſlower, it will give more time to any waters ſituated on one ſide of it, to flow one way; and on this account alſo, the apparent agitation of them will be increaſed.

SECT. II.

73. We are told, that, in the Liſbon earthquake of 1755, "the bar [at the mouth of the Tagus] was "ſeen dry from ſhore to ſhore; then ſuddenly the ſea,

* See art. 43.
† See art. 63. the note.

"like

" like a mountain, came rolling in; and about Bel-
" lem caſtle, the water roſe fifty feet almoſt in an
" inſtant; and, had it not been for the great bay
" oppoſite to the city, which received and ſpread
" the great flux, the low part of it muſt have been
" under water *." The ſame phænomena were obſerved to accompany the ſame earthquake at the iſland
of Madeira; where we are told, that, at the city of
Funchal, " the ſea, which was quite calm, was ob-
" ſerved to retire ſuddenly ſome paces; then riſing
" with a great ſwell, without the leaſt noiſe, and as
" ſuddenly advancing, it overflowed the ſhore, and
" entered the city. It roſe full fifteen feet perpen-
" dicular above high-water mark, although the tide,
" which ebbs and flows there ſeven feet, was then
" at half ebb. In the northern part of the iſland,
" the inundation was more violent, the ſea retiring
" there above one hundred paces at firſt, and ſud-
" denly returning, overflowed the ſhore, forcing
" open doors, breaking down the walls of ſeveral
" magazines and ſtorehouſes, and carrying away, in
" its receſs, a conſiderable quantity of grain, and
" ſome hundred pipes of wine †."

74. Both theſe appearances (which have been obſerved to attend ſeveral other earthquakes, as well as this) ſeem to admit of an eaſy ſolution, ſuppoſing the cauſe of them to lie under the bed of the ocean; for, in the farther progreſs of the communication between the fire and water, the vapour, that is

* See Hiſt. and Philoſ. of Earthq. p. 316.

† See Philoſ. Tranſ. vol. xlix. p. 432, *&c.*—or Hiſt. and Philoſ. of Earthq. p. 329.

gradually raiſed at firſt, will at laſt begin to raiſe the roof over the fire, which, being ſupported by ſo light a vapour, there will now be no want of fluidity in the matter it reſts upon, and the difference of ſpecific gravity between the two, inſtead of being ſmall, will be very great: hence, if any part of the roof gives way, it muſt immediately fall in, the vapour readily riſing, and taking its place; and a beginning being once made, a communication will be opened with numberleſs clefts and fiſſures, that muſt occaſion the falling in of vaſt quantities of matter, which, as ſoon as the vapour can paſs round them, will want their ſupport; then will follow the great * effects already deſcribed.

75. Now, whilſt the roof is raiſing, the waters of the ocean, lying over it, muſt retreat, and flow from thence every way; this, however, being brought about ſlowly, they will have time to retreat ſo gently, as to occaſion no great diſturbance: but as ſoon as ſome part of the roof falls in, the cold water contained in the fiſſures of it, mixing with the ſteam, will immediately produce a vacuum, in the ſame manner as the water injected into the cylinder of a ſteam engine, and the earth ſubſiding, and leaving a hollow place above, the waters will flow every way towards it, and cauſe a retreat of the ſea on all the ſhores round about: then preſently, the waters being again converted by the contact of the fire into vapour, together with all the additional quantity, which has now an open communication with it, the earth will be raiſed, and the waters over it will be made to flow every

* See art. 56 to 60 incluſive.

way,

way, and produce a great wave immediately succeeding the previous retreat *

SECT. III.

76. That great quantity of water, which we have supposed to be let out upon subterraneous fires, and, by that means, to produce earthquakes, will supply us with a reason, why they observe a sort of periodical return. This water must extinguish a great portion of the burning matter, in consequence of which, it will be contracted within much narrower bounds; and though the effects before described could not take place at first, but by the great extension of the heated matter, yet, after they have once taken place,

---

* It may, perhaps, be objected, that these phænomena may as easily be occasioned by a vapour generated under the dry land, which, by first raising the earth upon the sea-shore, would make the waters retreat; and that the return of them again, upon its subsiding into its place, might cause the subsequent wave. That this may be the case, in some instances, is not impossible, but, I believe, upon examining the particular circumstances, it will generally be found to be otherwise; and there cannot be any doubt about it, in the case of the Lisbon earthquake; for the retreat was observed to precede the wave, not only on the coast of Portugal, but also at the island of Madeira, and several other places: now, if the retreat had been caused by the raising of the earth on the coast of Portugal, the motion of the waters occasioned by this means, wnen propagated to Madeira, must have produced a wave there previous to the retreat, contrary to what happened; nor could the motion of the waters at Madeira be caused by the earthquake at that place. because it did not happen till above two hours after; whence it is manifest, that it must have been owing to the continuation of a motion propagated from the place, where the carthquake exerted its first efforts. And we may observe, in general, that this must always be the case, whenever the retreat does not happen till some considerable time after the earthquake.

 they

they may well continue to do ſo for ſome time; for the great diſturbance in the firſt inſtance, by the falling in of a great part of the roof, muſt render the frequent communication between the fire and water not only very eaſy, but almoſt unavoidable: and this will continue to be ſo, till the roof is well ſettled, and the ſurface of the melted matter ſufficiently cooled, after which, it may require a long time for the fire to heat it again ſo much, as will be neceſſary to make it produce the former effects. Now, as the matter has been more or leſs cooled, or as the combuſtible materials are with more or leſs difficulty ſet on fire again, as well as on account of other circumſtances, the returns of theſe effects will be later or earlier; but though they will not, for this reaſon, obſerve any exact period, yet they will generally fall within ſome ſort of limits, till either the matter that occaſions them is conſumed, (which, probably, will ſeldom happen in leſs than many ages) or till the fires open themſelves a paſſage, and become volcanos.

SECT. IV.

77. I have already intimated, that the moſt extenſive earthquakes frequently take their riſe from the ſea. According to the deſcription of the * ſtructure of the earth before given, any combuſtible ſtratum muſt lie at greater depths in places under the ocean, than elſewhere; hence far more extenſive fires may ſubſiſt there, than where the quantity of matter over them is leſs; for any vapour raiſed from ſuch fires,

* See art. 43.

having

having both a ſtronger roof over it, and being preſſed by a greater weight, (beſide the additional weight of the water) will not only be leſs at liberty to expand itſelf, and conſequently of leſs bulk, but it will alſo be eaſily driven away towards the parts round about, where the ſuperincumbent matter is leſs, and therefore lighter. On the other hand, any vapour raiſed from fires, where the ſuperincumbent matter is lighter, finding a weaker roof over it, and being not ſo eaſily driven away under ſtrata, that are thicker and heavier, will be very apt to break through, and open a mouth to a volcano; and it muſt neceſſarily do this long before the fires can have ſpread themſelves ſufficiently, to be near equal to thoſe which may ſubſiſt in places that lie deeper. All this ſeems to be greatly confirmed by the ſituation of volcanos, which are almoſt always found on the * tops of mountains, and thoſe often ſome of the higheſt in the world.

78. If, then, the largeſt fires are to be ſuppoſed to ſubſiſt under the ocean, it is no wonder that the

* Perhaps this may ſupply us with a hint (if the conjecture is not thought extravagant) concerning the manner in which theſe mountains have been raiſed, and why the ſtrata lie generally more inclining from the mountainous countries, than thoſe countries themſelves; an appearance not eaſily to be accounted for, but upon the ſuppoſition, that the upper parts of the earth reſt upon matter, in ſome degree, though not perfectly fluid, and that this matter is lighter than the earth that reſts upon it. This conjecture, however, will probably be thought leſs ſtrange, if it be conſidered, that the new iſlands, formed about Santerini and the Azores, have ſome of them been raiſed from 200 to 300 yards, and upwards; a height which might well enough intitle them to the denomination of mountains, if they had been raiſed from lands not lying under the ocean. [See Fig. 3.]

moſt

moſt extenſive earthquakes ſhould take their riſe from thence: the great earthquake of Liſbon has been * ſhewn to have done ſo; and that the cauſe of it was alſo at a greater depth, than that of many others, appears from the greater † velocity with which it was propagated.

79. The great earthquake that deſtroyed Lima and Callao in 1746, ſeems alſo to have come from the ſea; for ſeveral of the ports upon the coaſt were overwhelmed by a great wave, which did not arrive till four or five minutes after the earthquake began, and which was preceded by a ‡ retreat of the waters, as well as that at Liſbon. Againſt this, it may, perhaps, be alleged, that there were four || volcanos broke out ſuddenly, in the neighbouring mountains, when this earthquake happened, and that the fires of theſe might be the occaſion of it. This however, I think, is not very probable; for, to omit the argument of the wave, and previous retreat of the waters, already mentioned, it is not very likely, that more than one fire was concerned: beſides, the vapour, opening itſelf a paſſage at theſe places, could not well be ſuppoſed, if it took its riſe from thence, to ſpread itſelf far; eſpecially towards the ſea, where it is manifeſt,

* See art. 54. See alſo art. 94 to 97 incluſive.

† See the note to art. 63.

‡ Both the wave and previous retreat have been obſerved in the other great earthquakes, which have happened at Lima, and in the neighbouring country. See d'Ulloa's Voyage to Peru, part ii. book i. chap. 7.

|| If theſe volcanos were not new ones, but only old ones which broke out afreſh, [See the note to art. 34.] the argument will come with ſtill greater force.

that

that the ſtrata over it were of great thickneſs, as appears from the great velocity with which the earthquake was propagated there: the ſhocks alſo continued with equal, or nearly equal violence, for ſome months after the openings were made; whereas, if theſe fires had been the cauſe of them, they muſt immediately have ceaſed, upon the fires finding a vent, as it has happened in other * caſes. It is therefore much more probable, that a very large quantity of vapour, taking its riſe from ſome far more extenſive fire under the ſea, ſpread itſelf from thence; and as it paſſed in places, where the roof over it was naturally much thinner, as well as greatly weakened by the undermining of theſe fires, it opened itſelf a paſſage, and burſt forth.

80. As the moſt extenſive earthquakes generally proceed from the loweſt countries, but eſpecially from the ſea, ſo thoſe of a ſmaller extent are generally found amongſt the mountains: hence it almoſt always happens, that earthquakes, which are felt near the ſea, if at all violent, are felt alſo in the higher lands; whereas there are many amongſt the hills, and thoſe very violent ones, which never extend themſelves to the lower countries. Thus we are told, that, at Jamaica, " † ſhakes often happen in " the country, not felt at Port-Royal; and ſome- " times are felt by thoſe that live in and at the foot

* See art. 28.

† This is taken from an account of the earthquake that happened at Jamaica in the year 1692, which, as well as ſome others before-mentioned, was attended with the wave and previous retreat. See Philoſ. Tranſ. N° 209. or Lowthorp's Abr. vol. ii. p. 417 and 418.

" of the mountains, and by no body elſe." On the other hand, the earthquake that deſtroyed Port-Royal extended itſelf all over the iſland: and the ſame was obſerved of a ſmaller earthquake, that happened there in 1687-8; which latter undoubtedly came from the ſea, as appears by Sir * Hans Sloane's account of it.

81. Earthquakes of ſmall extent are alſo very common amongſt the mountains of Peru and Chili. Antonio d'Ulloa ſays, " Whilſt we were preparing for " our departure from the mountain Chichi-Choco, " there was an earthquake which was felt four " leagues round about: our field tent was toſſed to " and fro by it, and the earth had a motion like " that of waves; this earthquake, however, was " one of the ſmalleſt, that commonly happen in that " country." The ſame author tells us, in another place, that, " during his ſtay at the city of Quito, " or in the neighbourhood of it, there were two " earthquakes, violent enough to overturn ſome " houſes in the country, which buried ſeveral perſons " under their ruins."

SECT. V.

82. It is generally found, that earthquakes in hilly countries, are much more violent than thoſe, which happen elſewhere; and this is obſerved to be the caſe, as well when they take their riſe from the lower countries, as amongſt the hills themſelves. This appearance being ſo eaſily to be accounted for, from the ſtructure of the earth already deſcribed, I

* See Phil. Tranſ. N° 209, or Lowthorp's Abr. vol. ii. p. 410.

ſhall

ſhall content myſelf with eſtabliſhing the certainty of a fact, which tends ſo greatly to confirm it.

83. The earthquakes that have infeſted ſome of the towns in the neighbourhood of Quito, have not only been incomparably more violent than that which deſtroyed Liſbon, but they ſeem to have exceeded that alſo which deſtroyed Lima and Callao. In * Liſbon, many of the houſes were left ſtanding, although few of them were leſs than four or five ſtories high. At Lima alſo, it is only ſaid, that " all " the buildings, great and ſmall, or at leaſt the " greateſt part of them, were deſtroyed." Callao likewiſe, as it appears from the accounts we have of it, had many houſes left unhurt by the earthquake, till the wave came, which overwhelmed the whole town, and threw down every thing that lay in its way. All theſe effects ſeem to be greatly ſhort of thoſe produced by an earthquake that happened at Latacunga, in the year 1698, when the whole town, conſiſting of more than ſix hundred houſes, was entirely deſtroyed in leſs than three minutes time, a part of one only eſcaping; notwithſtanding that the houſes there are never built more than one ſtory high, in order, if poſſible, to avoid theſe dangers. Ambato, a village about the ſame ſize as Lataeunga, together with a great part of Riobamba, another town in the ſame neighbourhood, were alſo entirely deſtroyed by the ſame earthquake, and ſome others were either deſtroyed, or received conſiderable damage

* See Philoſ. Tranſ. vol. xlix. p. 403. where it is ſaid, " of " the dwelling-houſes, there might be about one fourth of them " that tumbled."

from it. At the ſame time, a volcano burſt out ſuddenly in the neighbouring mountain of Carguayraſo, as before-mentioned; and, "near Ambato, the earth "opened itſelf in ſeveral places, and there yet re- "mains, to the ſouth of that town, a cleft of four "or five feet broad, and about a league in length, "lying north and ſouth; there are alſo ſeveral other "like clefts on the other ſide of the river." The city of * Quito was affected at the ſame time, but received no damage, though it is no more than forty-two geographical miles from Latacunga, not far from whence the greateſt violence of the ſhock ſeems to have exerted itſelf. Theſe towns are ſuppoſed to ſtand by far the higheſt of any in the world, being as high above the level of the ſea, as the tops of ſome of the higheſt mountains in Europe; and the ground upon which Riobamba ſtands, wants but † ninety yards of being three times as high as Snowdon, the higheſt mountain in Wales.

84. The country upon which theſe towns ſtand, ſerves as a baſe, from whence ariſe another ſet of high lands and mountains, which are much the higheſt in the known world. Amongſt theſe mountains there are no leſs than ſix volcanos, if not more,

* The city of Quito ſtands lower than the level of Riobamba, by about 500 yards perpendicular. Though it eſcaped this, it has lately, however, been deſtroyed by another violent earthquake, that happened on the 28th April 1756, of which I have not yet ſeen any other particulars worth notice.

† This is according to Antonio d'Ulloa's account; but Monſ. Condamine makes it exactly three times the height of Snowdon, computing it at 1770 toiſes. [See his meaſure of a degree of the meridian.]

within

within an extent of 120 miles long, and leſs than thirty broad, the loweſt of which exceeds the height of Riobamba by above two thirds of a mile, and the higheſt by more than twice that quantity. Now, as the earthquakes have been more violent at the foot of theſe mountains, than in the lower lands, ſo they have been ſtill more violent towards the tops of them: this is ſufficiently manifeſt, from the many * rents made in them, and the rocks that have been broken off from them, upon ſuch occaſions: but it appears ſtill more manifeſtly, and beyond all diſpute, in the burſting forth of volcanos, which are almoſt always at the very † ſummit of the mountains, where they are found. In theſe inſtances, the earth, ſtones, &c. which lay over the fire, are generally ſcattered by the violence of the vapour, that breaks its way through, to the diſtance of ſome miles round about.

85. The great earthquake of the 1ſt November 1755, was alſo more violent amongſt the mountains, than at the city of Liſbon. We are told, that "the "mountains of Arrabida, Eſtrella, Julio, Marvan, "and Cintra, being ſome of the largeſt in Portugal, "were impetuouſly ſhaken, as it were, from their "very foundations; and moſt of them opened at "their ſummits, ſplit and rent in a wonderful man-"ner, and huge maſſes of them were thrown down "into the ſubjacent vallies ‡."

* See d'Ulloa's Voyage to Peru, part i. book vi. chap. 2.

† The only exceptions that I know of to this rule, are in thoſe caſes, where the higheſt part having an opening already, ſome freſh mouth opens itſelf in the ſide of the mountain.

‡ See Hiſt. and Philoſ. of Earthq. p. 317.

86. The ſame was obſerved at Jamaica likewiſe. In the earthquake that deſtroyed Port-Royal in 1692, we are told, that " more houſes were left ſtanding " at that town, than in all the iſland beſides. It was " ſo violent in other places, that people were violently " thrown down on the ground, where they lay with " their legs and arms ſpread out, to prevent being " tumbled about by the incredible motion of the " earth. It ſcarce left a planter's houſe or ſugar-" work ſtanding all over the iſland: I think it left " not a houſe ſtanding at Paſſage fort, and but one " in all Liganee, and none in St. Iago, except a few " low houſes, built by the wary Spaniards. In Cla-" rendon precinct, the earth gaped, and ſpouted up, " with a prodigious force, great quantities of water " into the air, twelve miles from the ſea; and all " over the iſland, there were abundance of openings " of the earth, many thouſands. But in the moun-" tains, are ſaid to be the moſt violent ſhakes of all; " and it is a generally received opinion, that the " nearer to the mountains, the greater the ſhake; " and that the cauſe thereof, whatever it is, lies " there. Indeed they are ſtrangely torn and rent, " eſpecially the blue, and other higheſt mountains, " which ſeem to be the greateſt ſufferers, and which, " during the time that the great ſhakes continued, " bellowed out prodigious loud noiſes and echo-" ings.

87. " Not far from Yallowes, a mountain, after " having made ſeveral moves, overwhelmed a whole " family, and a great part of a plantation, lying a " mile off; and a large high mountain near Port-

" morant,

" morant, near a day's journey over, is ſaid to be
" quite ſwallowed up.

88. " In the blue mountains, from whence came
" thoſe dreadful roarings, may reaſonably be ſup-
" poſed to be many ſtrange alterations of the like
" nature; but thoſe wild deſart places being very
" rarely, or never viſited by any body, we are yet
" ignorant of what happened there; but whereas
" they uſed to afford a fine green proſpect, now one
" half part of them, at leaſt, ſeem to be wholly de-
" prived of their natural verdure *."

SECT. VI.

89. I have ſuppoſed, that fires lying at the greateſt depths generally produce the moſt extenſive earthquakes, we muſt, however, except from this rule thoſe caſes where the depths are very great: for, as the weight of three miles perpendicular of common earth is capable of abſolutely repreſſing the vapour of inflamed gunpowder, ſo we may well ſuppoſe, that

---

* See Philoſ. Tranſ. N° 209. or Lowthorp's Abridg. vol. ii. p. 416, *&c.* where there is a great deal more to the ſame purpoſe. See alſo Hiſt. and Philoſ. of Earthq. p. 286 and 287.

From the authorities quoted in this ſection, it appears, how little reaſon there is for the notion, that either large cities, or towns ſituated near the ſea-coaſt, are more ſubject to violent earthquakes than others: it is not, however, much to be wondered at, that ſuch a notion ſhould have prevailed, after the great deſtruction that happened in ſo large and populous a city as Liſbon; ſince the demolition of a few ruinous houſes only, in ſuch a place, would have affected the imaginations of men more, and would have been more talked of, than the ſubverſion of whole mountains in ſome wild and deſart country, where at moſt half a dozen unknown ſhepherds might feel the effects of it, or perhaps only ſee it at a diſtance.

there may be a quantity of earth ſufficient to repreſs the vapour of water, and keep it within its original limits, though ever ſo much heated. Now, whenever this is the caſe, it is manifeſt, that it can produce no effect: or, it may happen, that though the quantity of earth may not be ſufficient abſolutely to repreſs the vapour, yet it may be ſo great, as to ſuffer it to expand but very little: in this caſe, an earthquake ariſing from it would be but of ſmall extent; the wave-like motion would be little or none; the vibratory motion would be felt every-where; and the propagation of the motion would be very quick. This laſt circumſtance being almoſt the only one, by which theſe earthquakes can be known from thoſe which owe their origin to ſhallower fires, it muſt be very difficult to diſtinguiſh them with certainty, as it is almoſt impoſſible to diſtinguiſh the difference of the time of their happening in different places, when the whole, perhaps, is comprehended within the ſpace of two or three minutes; poſſibly, however, ſome of the earthquakes, which we have had in England, may have been of this claſs.

SECT. VII.

90. If we would inquire into the place of the origin of any particular earthquake, we have the following grounds to go upon.

91. *Firſt*, The different directions, in which it arrives at ſeveral diſtant places: if lines be drawn in theſe directions, the place of their common interſection muſt be nearly the place ſought: but this is liable to great difficulties; for there muſt neceſſarily be great uncertainty in obſervations, which cannot, at

beſt,

beſt, be made with any great preciſion, and which are generally made by minds too little at eaſe to be nice obſervers of what paſſes; moreover, the directions themſelves may be ſomewhat varied, by the inequalities in the weight of the ſuperincumbent matter, under which the vapour paſſes, as well as by other cauſes.

92. *Secondly*, We may form ſome judgment concerning the place of the origin of a particular earthquake, from the time of its arrival at different places; but this alſo is liable to great difficulties. In both theſe methods, however, we may come to a much greater degree of exactneſs, by taking a medium amongſt a variety of accounts, as they are related by different obſervers. But,

93. *Thirdly*, We may come to the greateſt degree of exactneſs in thoſe caſes, where earthquakes have their ſource from under the ocean; for, in theſe inſtances, the proportional diſtance of different places from that ſource may be very nearly aſcertained, by the interval between the earthquake and the ſucceeding wave: and this is the more to be depended on, as people are much leſs likely to be miſtaken in determining the time between two events, which follow one another at a ſmall interval, than in obſerving the preciſe time of the happening of ſome ſingle event.

94. Let us now, by way of example, endeavour to inquire into the ſituation of the cauſe, that gave riſe to the earthquake of the 1ſt of November 1755, the place of which ſeems to have been under the ocean, ſomewhere between the latitudes of Liſbon and Oporto, (though probably ſomewhat nearer to the

the former) and at the diſtance, perhaps, of ten or fifteen leagues from the coaſt. For,

95. *Firſt*, The direction, in which the earthquake arrived at Liſbon, was from the north-weſt; at Madeira it came from the north-eaſt; and in England it came from the ſouth-weſt; all of which perfectly agree with the place aſſumed *

96. *Secondly*, The times in which the earthquake arrived at different places, agree perfectly well alſo with the ſame point. And,

97. *Thirdly*, The interval between theſe, and the time of the arrival of the ſubſequent wave, concur in confirming it. That all this might appear the better, I have ſubjoined the following table, aſſuming the point, from whence I compute, at the diſtance of about a degree of a great circle from Liſbon, and a degree and half from Oporto. In conſequence of this ſuppoſition, I have added three minutes to the interval between the time when the ſhock was felt at Liſbon, and at the ſeveral other places. The firſt column in the table contains the names of places; the ſecond, the diſtances from the aſſumed point, reckoned in half degrees; the third, the time that the earthquake took up in travelling to each, expreſſed in minutes; and the fourth contains the time in which the wave was propagated, from its ſource to the reſpective places, expreſſed in minutes likewiſe.

* All theſe directions, together with the times when the earthquake, as well as the ſucceeding wave, arrived at different places, (two or three only excepted) are taken from the 49th volume of the Philoſ. Tranſ. and the Hiſt. and Philoſ. of Earthq. To theſe, I muſt refer the reader for the particular authorities, which, as they are very numerous, I was not willing to quote at length.

Liſbon

	Half deg.	Min.	Min.
Lifbon * - - - - - - - -	2	3	12
Oporto * - - - - - - -	3	5	
Ayamonte - - - - - - -	6		53
Cadiz - - - - - - - -	9	12	82
Madrid - - - - - - - -	9	11	
Gibraltar - - - - - - -	11	18	
Madeira - - - - - - - -	19	25	152
Mountfbay - - - - - - -	20		267
Plymouth - - - - - - -	21		360
Portfmouth - - - - - - -	23	29	
Kingfale - - - - - - -	23		290
Swanfea - - - - - - - -	24		530
The Hague - - - - - - -	30	32	
Lochnefs - - - - - - -	33	66	
Antigua - - - - - - - -	98		565
Barbadoes - - - - - - -	101		485

98. In computing the times in the preceding table, allowance was made for the difference of longitude, as it is laid down in the common maps, which are

* It appears, by all the accounts, that the interval between the earthquake and wave, either at Oporto or Lifbon, was not long: I have met with no account yet, however, which tells us how long it was at the former, and only one which mentions it at the latter, where it is faid to have been nine minutes. [See *Memoires fur les tremblemens de Terre, p.* 245. compared with Hift. of Earthquakes, p. 315.] Thefe intervals, if we knew them exactly, might have ferved, perhaps, to afcertain the diftance of thofe two places from the original fource a little more accurately; but, as the diftance of neither from thence could be very great, a fmall difference in them would hardly fenfibly affect any of the others; from which, therefore, we may draw the fame general conclufions, as if they were exact.

not always greatly to be depended on. The times themſelves alſo are often ſo careleſly obſerved, as well as vaguely related, that they are many of them ſubject to conſiderable errors; the concurrent teſtimonies, however, are ſo many, that there can be no doubt about the main point; and, that the errors might be as ſmall as poſſible, I have not only endeavoured to ſelect thoſe accounts that had the greateſt appearance of accuracy, but, in all caſes where it was to be had, I have always taken a mean amongſt them. In many of the accounts, the relaters ſay only between ſuch hours, or about ſuch an hour: of this kind were the accounts of the times of the agitation of the waters at The Hague and Lochneſs, which vary the moſt from a medium of the reſt, the former erring about ſeven minutes in defect, and the latter about twenty minutes in exceſs; with regard to the latter, however, I muſt obſerve, that, from the account itſelf, it is probable the agitation happened ſooner than eleven o'clock, which is the time mentioned. The accounts alſo of the time of the agitation of the waters in the northern parts of England, ſeem to confirm the ſame thing *.

99. It is obſervable, in the preceding table, that the times, which the wave took up in travelling, are

* As the ſhorteſt way that the vapour could paſs from near Liſbon to Lochneſs was under the ocean, poſſibly it might, on that account, be ſomewhat retarded; for the water adding to the weight of the ſuperincumbent maſs, and not to its elaſticity, muſt produce this effect in ſome degree; it is probable, however, that this could make no great difference, as the motion ſeems to have been very little retarded in its paſſage from the original ſource to Madeira, to which place, I ſuppoſe, it muſt have paſſed under deeper ſeas than would be found in its road to Scotland.

not

not in the ſame proportion with the diſtances of the reſpective places from the ſuppoſed ſource of the motion; this, however, is no objection againſt the point aſſumed, ſince it is manifeſt, wherever it was, that it could not be far from Liſbon, as well becauſe the wave arrived there ſo very ſoon after the earthquake, as becauſe it was ſo great, riſing, as we are told, at the diſtance of three miles from Liſbon, to the height of fifty or ſixty feet. The true reaſon of this diſproportion, ſeems to be the difference in the depth of the water; for, in every inſtance in the above table, the time will be found to be proportionably ſhorter or longer, as the water through which the wave paſſed was * deeper or ſhallower. Thus the motion of the wave to Kingſale or Mountſbay (through waters not deeper in general than 200 fathoms) was ſlower than that to Madeira, (where the waters are much deeper) in the proportion of about three to five; and it was ſlower than that to Barbadoes, (where its courſe lay through the deepeſt part of the Atlantic ocean) nearly in the proportion of one to three: ſo likewiſe the motion of it from the Scilly iſlands to Swanſea in Wales (where the depth gradually diminiſhes from about ſixty or ſeventy fathoms to a very ſmall matter) was ſtill ſlower than that to Kingſale, in the proportion of leſs than one to three: the ſame thing is obſervable with regard to

* We have an inſtance to this purpoſe in the tides, which, in deep waters, move with a velocity that would carry them round the whole earth in a ſingle day; but as they get into ſhallower waters, they are greatly retarded: and we are told, that in the river of Amazons, the ſame tide is found running up to the tenth or twelfth day, before it is entirely ſpent. [See *Condamine's Voyage down the Maranon.*]

 Plymouth

Plymouth alſo, where the wave arrived about ninety minutes later than at Mountſbay, though the difference of their diſtance from the firſt ſource could not, upon any ſuppoſition, be more than forty or fifty miles.

Sect. VIII.

100. If we would inquire into the depth, at which the cauſe lies, that occaſions any particular earthquake, I know of no method of determining it, which does not require obſervations not yet to be had; but if ſuch could be procured, and they were made with ſufficient accuracy, I think ſome kind of gueſs might be formed concerning it: for,

101. *Firſt*, In thoſe inſtances, where the vapour diſcharges itſelf at the mouths of volcanos, (as in the caſe of the earthquake at Lima) it might, perhaps, be poſſible for a careful obſerver to trace the * thickneſs of the ſeveral ſtrata from thence to the place where the earthquake took its riſe, or at leaſt as far as the ſhore, if it took its riſe from under the ſea. If this could be once done in any one inſtance, and the velocity of ſuch an earthquake nicely determined, we might then gueſs at the depth of the cauſe in other earthquakes, where we knew their velocity, by taking the † depths proportional to thoſe velocities, which probably would anſwer very nearly.

102. *Secondly*, If, in any inſtance, it ſhould be poſſible to know how much the motion of any earthquake was retarded by paſſing under the ocean, the

* This is upon the ſuppoſition, that the under ſtrata, in aſcending up the hills, come to the day in the manner before deſcribed. See art. 43. and Fig. 3.

† See the note to art. 63.

depth

depth of the ocean being known, the depth at which the vapour paſſed would be known alſo; for the velocity under the water would be to the velocity, if there had been no water, in the ſubduplicate ratio of the weight in the latter caſe to the weight in the former; hence allowing earth to be about two and half times the weight of water, the depth will be readily found.

103. *Thirdly*, Let us conceive the earth to be formed according to the idea before given of it, and that the ſame ſtrata are at a medium of the ſame thickneſs for a very great extent, as well in thoſe places, where ſeveral of the upper ones are wanting, as where they are not. Upon this ſuppoſition, we may diſcover the depth, at which the vapour paſſes, by comparing the ſeveral velocities of the ſame earthquake in places, where the * thickneſſes of the ſuperincumbent maſs are different. It muſt be acknowleged, indeed, that ſuch obſervations with regard to time, as would enable us to determine theſe velocities, are in general much too nice to be expected: the matter, however, is not altogether deſperate, as we may collect them, in ſome meaſure perhaps, from other circumſtances, ſuch, for inſtance, as the degree of † agitation in different waters, the proportional ‡ ſuddenneſs, with which the earth is lifted in different places, *&c.*

104. As

* In order to know this difference, it will be neceſſary to trace the thickneſs of thoſe ſtrata, which are found in ſome of the places, but are wanting in others.

† See art. 71 and 72.

‡ This may be known from the diſtance to which the mercury ſubſides in the barometer, upon the firſt raiſing of the earth by the vapour.

104. As the obſervations relating to the earthquake of the 1ſt of November 1755 are too groſs, it would be in vain to attempt, by any of the foregoing methods, to determine with any certainty the depth at which the cauſe of it lay; but, if I might be allowed to form a random gueſs about it, I ſhould ſuppoſe, (upon a compariſon of all circumſtances) that it could not be much leſs than a mile, or a mile and half, and I think it is probable, it did not exceed three miles.

CONCLUSION.

105. Thus have I endeavoured to ſhew how the principal phænomena of earthquakes may be produced, by a cauſe with which none, that I have ſeen, appear to me to be incompatible. As I have not knowingly miſrepreſented any fact, ſo neither have I deſignedly omitted any that appeared to affect the main queſtion; but, that I might not unneceſſarily ſwell what had already much exceeded the limits at firſt intended for it, I have omitted,

106. *Firſt*, Thoſe minuter appearances, which almoſt every reader would eaſily account for, from what has been ſaid already, and which did not ſeem to lead to any thing farther: ſuch, for inſtance, are the ſudden ſtopping and guſhing out of fountains, occaſioned by the opening or contracting of fiſſures; the dizzineſs and ſickneſs people feel, from the almoſt imperceptible wave-like motion, *&c.*

---

vapour. I don't find, that this phænomenon, which is a common attendant on earthquakes, was obſerved any-where, at the time of the earthquake of the 1ſt of November 1755, except at Amſterdam, where the mercury ſubſided more than an inch. See Hiſt. and Philoſ. of Earthq. p. 309.

107. *Secondly*, Thoſe appearances which ſeemed to depend upon particular circumſtances, and of which, therefore, unleſs we had a more exact knowlege of the countries where they happened, it would have been impoſſible to give any account, without having recourſe to uncertain conjectures; of this kind, was the greater agitation of the waters in the lakes of Switzerland, at the time of the earthquake of the 1ſt of November 1755, than during the * earthquake of the 9th of December following, though the houſes upon the borders of them were more violently ſhaken by the latter. And,

108. *Laſtly*, Thoſe appearances, which only ſeem to have an accidental connection with earthquakes, or the cauſes of them; of this kind, are the effects which, in ſome inſtances perhaps, they produce on the weather; the diſtempers which are ſometimes ſaid to ſucceed them; the diſturbance which, we are told, they have ſometimes occaſioned, during the ſhocks, in the direction of the magnetic needle, *&c.* none of which are obſerved to be conſtant attendants on earthquakes, nor do they ſeem materially to affect the ſolution given either one way or other.

---

* See Monſieur Bertrand's *Memoires ſur les tremblemens de Terre.*

F I N I S.

[illegible] *Secondly*, These appearances [illegible] figured to depend on a particular circumstance, and of which therefore, unless we had a more exact knowledge of the [illegible] it would have been very difficult to give any decision, without having recourse to uncertain conjectures; for this kind [illegible] in the lakes of Switzerland, at the time of the earthquake of [illegible] in September 1755, and during the earthquake of the [illegible] of December following, though the [illegible] [illegible] of them were [illegible] violently shaken [illegible] lakes [illegible]

La[illegible]

to [illegible] [illegible] [illegible] [illegible] the cause [illegible] [illegible] [illegible] perhaps [illegible] the [illegible] [illegible] [illegible] [illegible] [illegible] [illegible] [illegible] have [illegible] [illegible] direction [illegible] [illegible] [illegible] [illegible] earthquake not [illegible] [illegible] given [illegible]

[illegible]

Printed in the United States
By Bookmasters